"What is the proper balance between religious expression and science education in a liberal democracy? To be 'free,' citizens must express their deepest held convictions but also think critically and ascertain facts. Susan Liebell, in an unusually probing and utterly fearless fashion, explores this question, thoroughly reshaping our understanding of the relationship of religion and politics in the United States. *Democracy, Intelligent Design, and Evolution* prioritizes science over religion—but it does so in ways that will surprise 'secularists,' and likely will earn the grudging respect of politically active religionists. A must-read for anyone interested in civic education in our supposedly 'post-secular age.'"
                                —John P. McCormick, *University of Chicago*

"Susan Liebell has written a wonderful and extraordinarily timely book. Through an uncommonly rich mix of theory, law, and political analysis, she offers a thoughtful and penetrating analysis of the complex role that science plays in democratic politics and civic life. During a time in which science education is a subject of great political debate, we should pay close attention to Liebell's arguments. As a citizen you will benefit from reading this book."
                                —Jack Knight, *Duke University*

Democracy, Intelligent Design, and Evolution

Should alternatives to evolution be taught in American public schools or rejected as an establishment of religion? *Democracy, Intelligent Design, and Evolution* argues that accurate science education helps shape a democratic temperament. Rather than defending against Intelligent Design as religion, citizens should defend science education as crucial to three aspects of the democratic person: political citizenship, economic fitness, and moral choice. Through an examination of *Tammy Kitzmiller et al. v. Dover Area School District*, contemporary political theory, and foundational American texts this volume provides an alternative jurisprudence and political vocabulary urging American liberalism to embrace science for citizenship.

Susan P. Liebell is Assistant Professor of Political Science at Saint Joseph's University. She worked in New York state government before receiving her PhD from the University of Chicago. Her scholarship employs foundational texts to explore modern political problems.

Routledge Research in American Politics and Governance

Democracy, Intelligent Design, and Evolution

Science for Citizenship

Susan P. Liebell

Routledge
Taylor & Francis Group

LONDON AND NEW YORK

First published 2014
by Routledge
711 Third Avenue, New York, NY 10017

Simultaneously published in the UK
by Routledge
2 Park Square, Milton Park, Abingdon, Oxfordshire OX14 4RN

*Routledge is an imprint of the Taylor and Francis Group,
an informa business*

First issued in paperback 2015

Library of Congress Cataloging-in-Publication Data

Liebell, Susan P.
  Democracy, intelligent design, and evolution : science for citizenship / Susan P. Liebell.
    pages cm. — (Routledge research in American politics and governance ; 9)
  1. Kitzmiller, Tammy—Trials, litigation, etc.   2. Dover Area School
District (Dover, Pa.)—Trials, litigation, etc.   3. Evolution (Biology)—Study
and teaching—Law and legislation—Pennsylvania—Dover.   4. Intelligent
design (Teleology)—Study and teaching—Law and legislation—
Pennsylvania—Dover.   5. Citizenship—United States.
  I. Title.
    KF228.K589L54 2013
    344.748'41077—dc23
    2013005829

ISBN 978- 0-415-89765-5 (hbk)
ISBN 978- 1-138-99948-0 (pbk)
ISBN 978-0-203-72442-2 (ebk)

Typeset in Sabon
by Apex CoVantage, LLC

To Adam, Julia, and Eli

Contents

Figures

Acknowledgements

I have always loved the unexpected thoughts that emerge from a workshop. So, I am pleased to recall that the idea for this book emerged from a post workshop conversation with John McCormick at the University of Chicago. Presentations at George Washington and Duke shaped the argument and I am grateful for the hard questions—particularly those posed by Ingrid Creppell, Ruth Grant, Jack Knight, and Michael Lienesch. At Saint Joseph's University, I am privileged to have colleagues from art, education, history, sociology, management, philosophy, and political science who have enriched my work: Amber Abbas, Lisa Baglione, Jim Boettcher, Jay Carter, Melissa Chakars, Michael Clapper, Emily Hage, Christopher Close, Kaz Fukuoka, Betsy Linehan, Dennis McNally, Elizabeth Morgan, Rob Moore, Andrew Payne, Ian Petrie, and Regina Robson. In particular, Richard Warren's insightful comment about the *Plyler* case pushed me to think outside of the evolution jurisprudence.

Some material from chapters 1 and 3 appear with the permission of Cambridge University Press: Susan Liebell "Rethinking Dover: Religion, Science, and the Values of Citizenship," *Politics and Religion*, Volume 5, Issue 2, August 2012. Copyright © Religion and Politics Section of the American Political Science Association 2012. Special thanks to Tom Davis at Bryan College for use of their photo archive.

Saint Joseph's University provided a summer research grant and support for conference presentations at the American, Western, and Midwest Political Science Associations as well as the Association for Political Theory. The thoughtful scholars who raised questions in those venues all contributed to this work. Aubrey Bellezza and Emily Reineberg provided editorial assistance. Matthew Mano assisted with research on state laws and curriculum and Meghan Sprout updated data on voting and education.

Phil Schatz, Tom Kane, Rich Warren, and Lisa Baglione traveled beyond collegiality to keep me working on this book. Maria Kefalas and Patrick Carr opened their home providing laughter, food and a critical legal reference. Ann Davies, Brad Evans, Sophie Evans, Jim Johnson, Steven Laymon, and Randy Petilos—since cold times in Hyde Park—have kept me intellectually honest. Paul McLean has, in and out of marriage, supported my work and gracefully endured the complexities of raising children while working in two different states.

I am deeply thankful to friends who have supported me—emotionally and intellectually—throughout the years: Martha Martin Reese provided insightful comments, wit, and the fish graphic; Dara Botvinick walked miles discussing the writing process; Christine Perney asked good questions and also knew when not to ask; Lauren Anderson and Theresa Bergherr listened at crucial moments; Alicia Jackson made me smarter with her questions and comments; and Scarlett McCahill worked editorial magic in the end-game. Jay Carter's friendship, conversation, affection, editorial help, and chili have been a source of strength. From my dearest friend Tory Ferrera, I have received editing suggestions, insight into the writing process, and, most important, decades of precious friendship.

My parents, Barbara and Edward Liebell, raised me to ask questions and care about nature, politics, music, and art. My wonderful cousins, Amy and Jonathan Fleming, have offered ideas, enthusiasm, and their Cape house for a work retreat.

It is impossible to express what I owe to three generous women. Without Jane Mansbridge's intelligence, generosity, and patience, I would have never finished my first project. Ruth Grant pulled me aside, just before the birth of my twins, to offer advice on writing and mothering that proved essential and she has supported me at critical junctures. Engaging with Cristina Beltrán—in cafes, libraries, on walks, and at conferences—has made everything that I write better.

Eli asks great questions and supplied sources regarding science and politics. Adam never fails to offer help and weighed in on the title. Julia sits across the table from me like a colleague and thoughtfully asks about my progress. Through who they are and what they do, my three wonderful and different children remind me of what matters most.

Introduction
Monkey Trials, Monkey Bills, and the Politics of Political Identity

In spring of 1925, the state of Tennessee tried John Scopes for teaching evolution—rather than Genesis—in a public school. State of the art telegraph, telephone, newsreel, and radio microphones carried the courtroom drama to American as well as international audiences. Two papers from London covered the trial, and more words were transmitted to Britain about the Scopes trial than for any previous American event. Chicago radio station WGN's live trial coverage—costing $1,000 per day just for the phone lines—was a national first. The opposing attorneys were national celebrities. Clarence Darrow, a leading member of the American Civil Liberties Union (ACLU) and opponent of the death penalty, had successfully defended Leopold and Loeb in the "trial of the century" the previous year. William Jennings Bryan, three-time presidential candidate, WWI pacifist, trustbuster, and prohibitionist, was a celebrated orator and opponent of evolution. Darrow portrayed Scopes as a teacher whose individual rights were being thwarted by a state insisting on excluding modern science when it conflicted with the Bible. Bryan countered that evolution, a questionable theory full of guesses, endangered the moral beliefs of children. The state could and should forbid the teaching of material inimical to the public welfare. H. L. Mencken dubbed Dayton, Tennessee, "monkeytown" and depicted its citizens as "yokels" and "morons." He saw Bryan as a "buffoon" dishing out "theologic bilge." The spectacle included Darrow calling Bryan as a witness and a live chimpanzee dressed in a suit walking around town. After Scopes was convicted, the verdict was quickly overturned on a technicality.[1] Widely seen as a public confrontation between fundamentalism and modernism, the Scopes "monkey trial" marks the beginning of an evolving jurisprudence on the teaching of best available science in American public schools.

Almost 90 years later, in spring of 2012, the state of Tennessee passed a law dubbed the "Monkey Bill" by its critics and the "Academic Freedom Law" by its supporters. In order to "inform students about scientific evidence and to help students develop critical thinking skills necessary to become intelligent, productive, and scientifically informed citizens," all public administrators and educators must teach "scientific subjects" that "may cause debate and disputation," such as "biological evolution, the chemical

Figure I.1 Joe Mendi, performing chimpanzee, at the Scopes Trial, 1925 Bryan College Archives

origins of life, global warming, and human cloning" as "scientific *contro-versies.*" Republican Governor Bill Haslam allowed the bill to become law without his signature.[2]

A number of organizations opposed the legislation, including the National Association of Biology Teachers, the American Association for the Advancement of Science, the American Civil Liberties Union of Tennessee, the American Institute for Biological Sciences, the Knoxville News Sentinel, the Nashville Tennessean, the National Association of Geosciences Teachers, the National Earth Science Teachers Association, the Tennessee Science Teachers Association, and all eight Tennessee members of the National Academy of Sciences. The law insists that teaching scientific subjects as controversies does not promote religion. Yet, many of these groups believe that teaching evolution as a "scientific controversy" is a "gateway for the teaching of creationism and intelligent design in public schools."[3]

We should understand the Tennessee law as part of a century-old fight over fundamentalism, modernism, and American citizenship—begun at the Scopes trial. Since the Supreme Court excluded the teaching of creationism or creationist science as religion that violated the First Amendment—"Congress shall make no law respecting an establishment of religion"—religious literalists have sought constitutionally permissible ways to thwart evolution education and include accounts of the origins of human life that are compatible with a literal reading of Genesis. Crafted in response to the Supreme Court's ruling, Intelligent Design (ID) appears to be the perfect legal instrument for those who favor teaching creationism or criticizing evolution. ID neither names God as the designer nor offers Christian texts as evidence. There is one simple claim: humans are too complex to have evolved through natural selection. Only an intelligent designer could have created blood-clotting cascades or the human eye. A tiny minority of credentialed scientists declare ID to be *science* that should be taught in public schools.[4] A U.S. district court struck down the mention of Intelligent Design as a scientific alternative to evolution. *Kitzmiller v. Dover Area School District* (2005), the last federal precedent, eliminated ID from the science classroom, using the Establishment Clause of the First Amendment, on the grounds that it was religion, not science.

This book argues that this "get religious science out!" line of legal reasoning is the wrong one for democracy. It is a parry that defensively wards off religion yet allows the ID movement to control the vocabulary and dangerously blurs the most critical issues at stake. Evolution education does not serve a religious purpose. It serves a democratic one. While ID demands a form of divine truth not available in a democracy, teaching evolution as the best available science based on observation and scientific method helps develop the critical thinking skills necessary to a democratic temperament. Controversies over ID and evolution are best understood as a contest over the foundation of citizenship in the United States. Americans miss the mark if they only ask, "Is it tolerable in a multifaith society to mention Intelligent Design in a science classroom?" Instead, we need to investigate and stress the ways that scientific education creates and maintains liberal democratic political identity.

Based on works in public law and political theory, this book advocates a positive argument regarding how courses in science help students think like democratic citizens. This is a stronger position than a negative or defensive argument for excluding Intelligent Design as religion: teaching science is important for democracy. Advocating for science is far more important than explaining why Intelligent Design should be excluded as religion. Teaching science and scientific method exposes students—future citizens, workers, and moral decision makers—to critical thinking skills essential to the maintenance and stability of a liberal democratic polity. Science education—particularly training in the scientific method—helps create a liberal civic identity that stresses thinking systematically, observing causality, and interrogating and revising traditional institutions and norms. This democratic temperament—which accepts and emphasizes critical thinking in politics, economics, and personal life—is enhanced when students are exposed to best available science, the scientific method, and facts about the natural world. Science is an essential and fundamental component of a democratic education that helps cultivate three constituent parts of the liberal individual: political citizenship, economic fitness, and moral personhood. More specifically, science education activates many of the rights enumerated in the Constitution. Without education, the right to vote or speak freely cannot empower the individual to express herself or protect the nation against tyrannical government.

Emphasizing science education does not make me an apologist for science as an "objective" arbiter of political problems. Science, particularly the theory of natural selection, has rationalized nefarious social, economic, and political movements such as eugenics or social Darwinism. Science is neither objective nor neutral—and it cannot claim to answer political, economic, or social questions by appealing to scientific method. But legitimate concerns about fetishizing science—like those offered by Wolin or Euben—often lapse into a discomfort with *all* of science.[5] As theorists attack an idealized science that offers objective answers, they ignore those practices of science that are important for democracy: imagination, the value of experimentation, criticism of peers, the value of error and failure, and the capacity to think anew. Promoting accuracy in science education must be distinguished from fetishizing science or advocating positivism.

The ID movement presents evolution as *speculation* and state-sponsored atheism. ID manipulates the vocabularies of science and liberal democracy to create a powerful conflict narrative. Educators who oppose specious science are cast as opposed to free discussion and godless. This book refutes this fallacious conflict narrative and reframes the discussion in terms of democratic politics. Schools do not eliminate baseless science to shut down inquiry or thwart belief in God. The best available science should be taught to encourage the critical temperament that supports democratic practices and institutions—regardless of personal religious beliefs. This book provides an alternative jurisprudence and theoretical vocabulary to the "religion out" approach based on Supreme Court precedents and American political thought from the eighteenth to the twentieth centuries. Science education is an integral part of self-governance—how we collectively govern the nation as citizens as well as how we govern ourselves as individuals in the economy and private life.

* * *

Throughout the last 10 years, evolution education has been undermined by state laws, practices, and curriculum requirements. Many states have considered legislation to limit the teaching of evolution, placed warning stickers regarding evolution on textbooks, or encouraged teachers to include Intelligent Design or creation-science (Arkansas, Florida, Indiana, Iowa, Kentucky, Louisiana, Maryland, Michigan, Mississippi, Missouri, New Hampshire, New Mexico, New York, Oklahoma, South Carolina, Texas, and Utah). Several states shored up evolution education, ruled in favor of keeping evolution in the curriculum or textbooks, or reversed earlier anti-evolution practices (California, Georgia, Florida, Kansas, Ohio, Texas). In some schools, evolution is not taught until high school (Indiana, Iowa, Kansas, Kentucky, Michigan, and Nebraska), listed as voluntary, or left unassisted (Missouri, Tennessee, Maryland). Some states single out evolution as a "theory" and require students to consider strengths and weaknesses (Colorado, Missouri, Montana, and West Virginia). Even New York—which has a well-regarded life science curriculum—tells students that "many" scientists believe biological evolution occurs through natural selection. Pennsylvania previously included *human* evolution but omitted the term in the most recent state standard. Only Florida, New Hampshire, Iowa, and Rhode Island unflinchingly stipulate human evolution in state curricula.[6]

These changes in state laws and curricula come at a time when American students are failing to excel in science. National assessments show only 21 percent of American 12th graders are proficient in science. International comparisons rank American students 23rd out of 65 or 11th out of 48 nations in science education.[7] The 2012 Fordham Institute evaluation of state science curriculum standards identified undermining evolution and failure to integrate scientific inquiry (along with vagueness in standards and a lack of the math that is integral to science) as two of the four main problems in American science curriculum standards.[8]

Ironically, Intelligent Design has successfully argued for laws and curriculum to support baseless science by manipulating two powerful narratives: science and democratic deliberation and rights.

The language of science—developed over hundreds of years—defines a theory as a "comprehensive explanation of some aspect of nature supported by a vast body of evidence."[9] Even established scientific theories are subject to continuing refinement based on new data or the development of better technologies for observation—like DNA analysis or mapping the genome. Although there is little chance that new evidence will show that the planets do *not* travel around the sun or that living things are *not* made of cells, scientists refers to heliocentric or cell *theory*. Like these other foundational scientific theories, the *theory* of evolution is supported by so much evidence and confirming experiments over a long period of time that it is assumed to be true, yet—like cell theory or plate tectonics—science will always refer to evolution as a *theory*.[10]

Anti-evolution advocates pervert the scientific term *theory* to their political advantage by exploiting the everyday meaning of a theory as unduly speculative or a hunch. All of the critical inquiry statutes as well as the textbook stickers insist that evolution is *merely* a theory to justify examining weaknesses or exploring alternatives. The sticker that triggered a federal court battle in Georgia is typical: "this textbook contains material on evolution. Evolution is a theory, not a fact, regarding the origin of living things. This material should be approached with an open mind, studied carefully, and critically considered."[11] There are no such warnings for cell theory or quantum physics. Essential elements of democratic discourse—critical inquiry, toleration of pluralism, freedom of discussion—are appropriated to introduce discredited theories focused on a designing deity.

These stickers and statutes not only play on prevailing understandings of the word *theory*. They also manipulate the history of science to suggest that Intelligent Design—like Galileo's heliocentrism—is a paradigm-shifting approach that is being suppressed by an entrenched and jealous scientific establishment.[12] The National Academy of Sciences, the American Association for the Advancement of Science, and scores of scientists have demonstrated the fallacies of Intelligent Design as well as ID's claim to have "peer reviewed" data.[13] Although 99.85 percent of scientists in the field believe in evolution, ID presents the tiny minority as potential Galileos or Einsteins. By simultaneously misrepresenting the meaning of theory (to suggest evolution is too speculative) and focusing on science's fundamental assumption of falsifiability, ID implies that it provides valuable scientific information.

In a similar manner, ID distorts a second essential narrative: democratic discussion and rights. The critical inquiry laws in Tennessee and Louisiana—and the ID movement generally—use the language of American liberal democracy: a mix of individual rights and the rule of law (liberalism) and majority rule (democracy) that supports freedom of speech, press, and religion in the name of individual choice, collective discussion, and self-rule. The law claims to protect teachers from censorship so they may present data relevant to "scientific controversies" and thereby encourage students to be inquisitive, use evidence, and "respond appropriately and respectfully to differences of opinion about controversial issues."[14] Questioning the science of evolution is presented as an extension of individual rights of freedom of speech and necessary for enhancing the critical skills of students. Why wouldn't all Americans support uncensored discussion in the classroom on controversial scientific subjects? The answer lies within the law's language: teaching scientific subjects as controversies does not "promote any religious or non-religious doctrine."[15] Courts have thrown out creation-science or Intelligent Design based on the state or school district's *intent* to inject Christian religion into the science classroom. Tennessee identifies secular reasons—to thwart any challenge of legislative motivation—yet the religious disclaimer is necessary given the history of *religious* challenge to evolution in the United States.

Some Christians believe in a literal interpretation of Genesis: God created the earth in 6 days approximately 6,000–10,000 years ago. Modern science demonstrates that the earth is more than 4 billion years old. For most world religions, evolution and faith in God are compatible, yet a minority of literalists have created a powerful narrative in which promoting evolution means *opposing* Christianity.[16] Since the Scopes trial, religious literalists have contended that evolution is state-sponsored atheism because evolution may encourage students to view the Bible as a historical document rather than the literal word of God. Despite Tennessee's disclaimer, teaching evolution as controversial promotes one set of beliefs: the young earth creationism of Genesis literalists.[17] Although Tennessee's law uses the languages of science and of democracy, it undermines the separation of church and state and limits the freedom of speech of teachers who support established scientific theories such as evolution. By presenting discredited science as "freedom of speech," the law encourages multiple voices at the expense of standards of scientific evidence—privileging religion rather than unheard scientific evidence.

Science uses the language of facts and truth—but the pseudo-scientific language of ID has won the day in many states and school districts. Those who oppose science and democracy have claimed the democratic vocabulary, and those who see science as part of democratic education need to take the language of democracy back. Intelligent Design is not democracy. It is either creationism wearing a lab coat or a dangerous form of post-modernism arguing for multiple truths and the imperfection of all theories of knowledge.

The success of ID and critical inquiry movements is more understandable if we consider the religious beliefs of Americans as well as their complex national origins. Even though Americans support the First Amendment's separation of church and state, they support religion in their private lives more than any other Western people. Unlike the strong majorities of Europeans who accept evolution as best available science, Americans believe God created fully formed humans (46%), or humans evolved with God's guidance (32%). Only 15 percent of Americans believe that humans evolved without the assistance of God.[18] The American preference for teaching both evolution and creationism may reflect the nation's compound origins. As Tocqueville insisted, the United States has two founding traditions: one rooted in Enlightenment reason and the other linked to English Protestantism.[19] The U.S. Constitution makes no reference to God as the source of rights, yet the Declaration of Independence speaks of rights endowed by our Creator.

Even within these two broad categories, there is infinite pluralism regarding citizenship. Those who favor an Enlightenment approach centered on a set of beliefs (the national creed)—rather than the religious origins—do not agree about the nature of personhood. Some, like Thomas Jefferson, believed that citizenship entails direct and genuine participation in politics. James Madison, on the other hand, preferred indirect participation through representation as sufficient for the preservation of democracy or the freedom of the individual. In the twenty-first century, these divisions remain—as well as conflict over the extent to which American liberalism should foster autonomy as a value. For some, public life represents the heart and soul of the government and the place where the individual can experience true freedom by giving herself laws to live by and transcending the politics of self-interest. For others, private life is the center of the moral, economic, and social world such that citizenship should be understood narrowly so as not to interfere in private or religious life. Those who ground the American experience in religious freedom and practice also widely diverge in what they want from citizenship. The Amish seek to live apart from mainstream institutions (e.g., social security or high school diplomas), while the Methodists understand their faith to compel them to engage issues of social justice and politics. These contradictions—the fact of pluralism, as John Rawls would have it—shape American political identity.[20] Yet identities are not static.

Religious opponents of evolution education understand that political systems create and maintain values through their education systems. These passers of monkey bills skillfully push Americans toward a particular understanding of the Constitution and science education—an understanding that supports *their* view of American citizenship rooted in Christian religion. In order to adjudicate plural views about education, American courts and contemporary political theorists depend upon an analysis of content: a categorical determination of what counts as religion and what counts as science. Defending against religion in the public schools obfuscates the real issue: why is evolution education important for democratic education?

This book provides a new legal and political vocabulary that enables an explicit justification of teaching science: science for self-governance and democratic temperament. Science education develops three aspects of the liberal person: political citizenship, economic fitness, and moral personhood. In order to reframe the debate over ID and evolution, this book moves between the present and the past.

Chapters one and two explore the tools we have at hand in American jurisprudence and political theory. Chapter one analyzes Supreme Court precedents and the *Dover* case to demonstrate that the Constitution pushes American courts to defensively exclude creationism rather than create a positive case for teaching the best available science. Chapter two explores what is at stake for liberalism by analyzing three liberal theories of education offered by Amy Gutmann, Harry Brighouse, and Stephen Macedo. Gutmann excludes creationism as a religious creed that threatens the quality of citizens' deliberation as equal citizens, but she does not clarify how science makes citizens more capable of public reason. A brief analysis of public opinion polls establishes the diversity of American beliefs as well as the lack of support for the scientific consensus that Gutmann depends upon to exclude creationism. Brighouse's call for education for autonomy cannot work given the need for liberalism to maintain more neutrality on ways of life, but his analysis reveals how certain *skills* contribute to public reason—and how science education might contribute to those citizenship capacities. Although Stephen Macedo's call for liberalism to grow a spine seems promising, his fear of positivism leaves him unable to protect science education. Although all three approaches fall short, elements of each lay the groundwork for an alternative liberal discourse to protect science based on skills for citizenship and a democratic temperament.

The next three chapters build the argument for science for citizenship. Using the education jurisprudence of the Supreme Court, Chapter 3 identifies *Plyler v. Doe* (1982) as a precedent for an alternative jurisprudence for science education. Rather than defend against creationism, the Court can insist that children have access to public education to become capable political, economic, and moral decision-makers. Chapter four joins the logic of *Plyler* with the insights of contemporary political theory as well as the works of John Dewey to present a skills-based approach to liberal education. Science education should be tied to the skills and substance that citizens need to be jurors, voters, and deliberating citizens. In the eighteenth century, John Adams, Thomas Jefferson, and Thomas Paine insisted that education was necessary to create the democratic temperament that would sustain popular government and actualize the rights of the individual. Chapter five demonstrates how the self-governing jurisprudence is embedded in early American political thought. Taken as a whole, the Supreme Court, contemporary democratic theorists, and early American thinkers help create a modern vocabulary for science education in a democratic liberal regime. The final chapter places Thomas Paine's understanding of science and religion as *compatible* in historical context. A brief review of how major Western religions have approached evolution demonstrates that the conflict narrative that Intelligent Design promotes is deeply misleading. The conclusion reflects on how it might be possible to build support for evolution education in today's politics.

Relying on the temperament of self-governance is both principled and politically practical. Given American pluralism, we should not demand education for a strong version of liberal autonomy. Given the political environment and the long-standing split in American political culture, a more modest claim—political and economic competence for self-governance—may produce a more radical outcome. No vision of citizenship supports passive, uneducated citizens unable to spot tyranny in their leaders ("to petition the government for a redress of grievances") or economic dependence.

Although the skills necessary to establish and maintain liberal democracy are notoriously difficult to specify, they include resisting tyranny, exercising rights (particularly those linked to dissent and criticism), and making independent choices in economic, personal, and political life. Education in a democracy should encourage critical, independent thinkers who can evaluate evidence, policies, candidates, culture, and traditions. Democratic people acting as citizens, jurors, consumers, soldiers, workers, or moral persons should be able to detect fallacies in arguments and, most important, be able to criticize and challenge authority.[21] At its best, a democratic temperament encourages the individual to imagine herself in the position of others in order to understand and tolerate other points of view and consider those points of view when making choices and policy.[22] All citizens will *not* be equally proficient, yet democracy requires a citizenry with some combination of these aptitudes to make decisions in public—as well as private—life.[23]

Economic fitness is related to this vision of the democratic citizen. Students must be educated to make choices in the economy—to seek, choose, hold, and change jobs. Again, there are public (national) and private (individual) reasons to cultivate individual economic capacity. Americans divide over optimal levels of growth or the importance of relative inequality in per capita income. Yet the nation is more stable when the economy (however defined) thrives and individuals are able to find and hold jobs. Individuals, regardless of their private goals and desires, need employment to fulfill basic humans needs (food, shelter, clothing), enjoy leisure, and (in some cases) experience work that is personally interesting or fulfilling.[24]

* * *

Whereas supporters of Intelligent Design recognize that they are fighting to establish education for a particular American political identity, supporters of evolution have failed to define and defend an explicitly secular and scientific liberalism. This book changes the dynamic of this one-way fight by supplying an alternative jurisprudence and a vocabulary that challenges the basic claims of the Intelligent Design movement and reframes the debate to emphasize the role of best available science in a liberal democracy. Few works in democratic education focus on the role of science in the development of a democratic temperament or see the Intelligent Design debate as significant to the formation of liberal political identity.[25] Though *science* is occasionally mentioned as part of a democratic education, *science education* is often ignored or confused with applied technology.[26] Martha Nussbaum's work on how the humanities further democratic education, for example, inappropriately conflates science education with technical or vocational instruction and often associates science with attention to applied technology, professionalization, and profit motives at the expense of critical inquiry. Nussbaum acknowledges that science and social science are crucial to democratic education but directs her inquiry to what is "both precious and profoundly endangered" because "nobody is suggesting leaving [science and social science] behind."[27] I disagree. Rather than assume the health of science education, we must carefully determine which *forms* of science education are—like elements of the humanities—profoundly endangered. This book does not outline a curriculum or pedagogical theory for developing the self-governing temperament. As I reframe the discussion, I leave the specifics of how students learn and how best to teach them to the appropriate scholars and practitioners.

Most books on democratic education subscribe to a dangerous assumption: the myth of an American consensus on standards of evaluation, particularly in science. Amy Gutmann, for example, assumes an American consensus regarding "secular standards of reasoning" such that "[m]ost Americans have reconciled the tenets of their faith with the findings of science." Creationism, she argues, can be rejected because the United States has "widespread acceptance" that "scientific standards of evidence and verification" should guide the content of knowledge transmitted to future generations.[28] Based on 30 years of polling data, no such consensus exists. This work provides an argument relevant to a uniquely fractured democratic polity in which religious literalists wield remarkable political and cultural power.

Of course, this book does not argue that science is the *only* means to a self-governing temperament. Many forms of education—including music, art, history, economics, literature, philosophy, theology, and more—contribute to citizenship, economic fitness, and moral personhood. My focus on science is justified by (1) the widespread attack on teaching evolution in state laws and school curricula (e.g., the language of the Tennessee law is not directed against supply side economics or restorative justice), (2) the importance of particular policy issues (e.g., global warming) that require evaluating scientific data and maintaining appropriate standards of evidence, and (3) the systematic ignoring of science education by scholars who value critical thinking and democratic deliberation.

This book is not a call for a more scientific politics or political theory. Using Dewey's ideas to encourage teaching best available science need not mean simplistic positivism. We must distinguish what science teaches us about the world (e.g., the age of fossils) from science dictating democratic politics. Understanding the evidence for and the dynamics of climate change *should* inform public policy, but climate scientists cannot—and should not—decide whether it is appropriate to ask developing countries to limit fossil fuels or developed nations to make transfer payments.

Liberals should not defensively exclude ID as established religion without providing positive arguments for secular standards and science. The defensive approach leaves liberalism at an impasse: a hollow legal victory in *Dover* that satisfies the supporters of science yet provides no defense against a religion-free ID that emphasizes academic freedom and "alternative views." Tennessee's new law creates the perfect storm: no mention of religion and an emphasis on "academic freedom." Advocates of ID have perverted the liberal language of democratic debate, and it is important to reclaim these important narratives. The liberal state depends upon science education that fosters critical thinking in order to reproduce critical liberal institutions and practices. We *can* have Intelligent Design or discussions of religion in biology classrooms, but there is a civic outcome that rests on making a choice and defending it. If we want best available science and scientific method instead, we have to admit that this is a political issue and engage it.

1 The Evolving Jurisprudence of the Supreme Court

Unpacking Dover and Intelligent Design

Dover is a predominantly Christian, white, and economically modest school district of 3,700 students just south of Harrisburg, Pennsylvania. In 2002, the president of the Dover School Board, Alan Bonsell, *publicly* declared two goals for the district—reinstate public prayer and teach creationism—even though the U.S. Supreme Court has ruled that both are unconstitutional establishments of religion. In 2003, Bonsell personally expressed his dissatisfaction with the teaching of evolution to the District's teachers. The district leadership and teachers clashed openly when the teachers requested a new biology textbook. The chair of the curriculum committee, William Buckingham, objected that the text extensively covered evolution and failed to adequately consider creationism. At a public meeting, several members of the school board supported the exclusion of the teacher-chosen textbook. Buckingham's wife (as a member of the public) spoke against offering anything but the Bible to the children and urged the audience to become born-again Christians. Alan Buckingham then challenged the audience to trace their ancestors to monkeys. He insisted that courts that have excluded the teaching of creationism take away the "rights of Christians." He concluded by asking the audience to stand up for Jesus.[1] In this religiously charged debate, preventing evolution education was central to the Board president's plan to return prayer and creationism to the public schools.

As the controversy brewed, Buckingham was contacted by the Discovery Institute—an institution at the forefront of the Intelligent Design movement—which supplies books, videos, and legal advice to school districts that encourage teaching ID as science. By 2004, Buckingham had discovered *Of Pandas and People*, an Intelligent Design work published by the Foundation for Thought and Ethics. An earlier version of the book used the term *creationism*, but, following the 1987 Supreme Court decision in *Edwards v. Aguillard* excluding creation-science from public schools, the Foundation meticulously edited the book: *design* replaced the word creationism.[2] Taken with ID as a possible approach for Dover, Buckingham made approval of the biology textbook contingent upon including *Of Pandas and People* as a reference work. Given the clear ruling in *Edwards v. Aguillard*, the board attorney expressed concern over a possible lawsuit.[3]

Amidst intense discussion in the community and among board members, 60 copies of *Panda* were anonymously donated to the high school. Later, testimony revealed that Buckingham had called for donations at his church, Harmony Grove Community Church, and Bonsell and Buckingham had worked to conceal the source.

Just as the books appeared in the library, the democratically elected school board passed a resolution (6 to 3) requiring teachers to caution their biology students about evolution:

> The Pennsylvania Academic Standards require students to learn about Darwin's theory of evolution and eventually to take a standardized test of which evolution is a part.
>
> Because Darwin's Theory is a theory, it is still being tested as new evidence is discovered. The Theory is not a fact. Gaps in the Theory exist for which there is no evidence. A theory is defined as a well-tested explanation that unifies a broad range of observations.
>
> Intelligent design is an explanation of the origin of life that differs from Darwin's view. The reference book, *Of Pandas and People*, is available for students to see if they would like to explore this view in an effort to gain an understanding of what intelligent design actually involves.
>
> As is true with any theory, students are encouraged to keep an open mind. The school leaves the discussion of the origins of life to individual students and their families. As a standards-driven district, class instruction focuses upon preparing students to achieve proficiency on standards-based assessments.[4]

The three dissenting school board members resigned. The teachers refused to read the statement, arguing that the statement violated their responsibilities as professional educators. Referring the students to *Pandas* as a "scientific resource" breached the teachers' "ethical obligation" to provide the students with "scientific knowledge that is supported by recognized scientific proof or theory."[5] The teachers cited a Pennsylvania law that forbids teachers from "knowingly and intentionally" misrepresenting subject matter or curriculum.[6] The community exploded with members attending meetings and writing letters. The two local newspapers received 225 letters to the editor and responded with 62 editorials. Sixty-five percent of the letters and 73 percent of the editorials framed the issue as religious.[7]

Pushing forward without the teachers, the *administrators* read the statement to the students.[8] In response, 11 families sued the district, insisting that the statement violated the First Amendment's ban on the establishment of religion. In the U.S. District Court, the families were represented by the American Civil Liberties Union (ACLU), Americans United for Separation of Church and State, and Eric Rothschild, a partner at Pepper Hamilton LLP and a member of the National Center for Science Education (NCSE) legal advisory. The Thomas More Law Center, a conservative Christian not-for-profit center committed to the religious freedom of Christians, represented the school district.

In *Tammy Kitzmiller, et al. v. Dover Area School District, et al.* (2005), the U.S. District Court ruled that ID was religion, not science, and must be excluded from the public school science curriculum. John E. Johns III, a conservative Republican appointed by President George W. Bush, wrote a lengthy and aggressive opinion excluding Intelligent Design. Even though God or Christianity was not mentioned in Intelligent Design materials, ID was creationism with a false scientific rhetoric. As such, ID must be excluded as religion using the Establishment clause of the First Amendment. When the school district held elections in 2005, none of the members who had voted for the Intelligent Design policy were reelected. The new board rejected the policy and refused to appeal the ruling to a higher court.

As part of the wider debate about religion in American life, the *Dover* case illustrates what has become a standard formula for adjudicating the separation of church and state: a categorical determination of what counts as religion. Occasionally, courts determine whether what is taught qualifies as science, but the Supreme Court has never accepted this approach. To win in *Dover*, the claimants had to successfully prove that ID was "religion." Although the Constitution encourages the courts to frame the issue as one of keeping religion *out* of public schools, the more important question remains: why does a liberal polity need to keep science *in*? By tracing the precedents relied upon in *Dover* and providing a critical evaluation of the *Dover* decision, this chapter demonstrates that the Constitution pushes the Court to provide a defensive strategy for the exclusion of creationism rather than creating a positive case for teaching evolution as best available science or the ability of potential citizens to distinguish between science that is good or bad. Although *Dover* fails to articulate how science education creates liberal democratic subjects, it provides valuable tools for crafting a new legal and theoretical approach.

THE EVOLUTION OF *EVOLUTION* IN AMERICA

Darwin published *The Origin of Species by Means of Natural Selection* in 1859. By the end of the century, evolution was selectively taught in U.S. public schools. Some courts have appealed to the history of evolution education to contextualize the "neutral" requests for creation-science or Intelligent Design in science classrooms. Opponents of evolution often stipulate that teaching evolution falsifies religion (William Jennings Bryan argued this position during the *Scopes* trial) or that evolution itself is a religious belief: atheism. Yet the *first* fights over Darwinian evolutionary theory did not pit science against religion in such a simple way. When Darwin introduced his theory of evolution, conservative—particularly Calvinist— theologians embraced it and found ways to place evolution *within* their theological frameworks. Liberal theologians struggled to square Darwin's theory (which emphasized the role of chance in natural selection) with their progressive conception of God, including an inevitable "material, social, and spiritual progress" in the world.[9] Yet anti-evolutionism and Christian *fundamentalism* are inextricably linked in the American case because anti-evolution played an integral part in a religious and political protest movement, beginning at the start of the twentieth century: Christian Fundamentalism. Evolution helped transform fundamentalism into a successful mass political movement, in part, because the teaching of evolution— first in colleges and later in secondary schools—could be easily identified and targeted. The authors of *The Fundamentals*—a massive publication— provided scientists—often discredited or unknown—to discredit evolution.[10] Although early Fundamentalist writers affirmed the compatibility of science and religion, later writers insisted that faith and reason were separate "spheres" and evolutionists were "increasingly defined as enemies of religion and society." Given the popularity of science in the early part of the century, anti-evolutionists tried to redefine science such that it came to be "linked in the public mind with biblical creation rather than evolution." These patterns repeat in contemporary debates over Intelligent Design, including the manipulation of scientific definitions of *theory* and *fact* as a means of discrediting evolution.[11]

Ultimately, Roman Catholicism, most forms of Judaism, and most denominations of Protestantism reconciled evolution and belief in a monotheistic god. Indeed, members of these groups have challenged laws mandating the teaching of creationism or filed *amicus curiae* briefs on behalf of parents. Although Christian literalist fundamentalists do not fully represent America's religiously faithful, a committed minority of literalists in the United States have long believed that teaching evolution denies biblical truth and weakens arguments for Divine creation. At the turn of the twentieth century, they organized to turn their views into legislation.

During the 1920s, 45 anti-evolution bills were introduced in 21 states and became law in Oklahoma, Florida, Tennessee, Mississippi, and Arkansas. In addition to legislative efforts, governors pushed for changes, including the censoring of textbooks.[12] In 1925, the state of Tennessee banned the teaching of "any theory that denies the story of the Divine Creation of man as taught in the Bible" in a public school. The ACLU located a teacher, John Scopes, who was willing to test the law.[13] Tennessee charged Scopes with teaching evolution.[14] In this first major legal challenge regarding science education, the court asked whether the state could ban material that conflicted with Genesis, *not* whether the state could teach creationism without violating the Establishment clause. The Supreme Court of Tennessee upheld Scopes conviction, and—most remarkably—the law remained in effect until 1967.

For roughly 35 years, states taught either evolution or creationism without a ruling from the U.S. Supreme Court. Although education is one of the responsibilities traditionally left to the states, such deference ended in 1968. In *Epperson v. Arkansas*, the Supreme Court unanimously struck down an Arkansas statute prohibiting the teaching of evolution in public schools. The Court found that the law attempted to "blot out a particular theory because of its supposed conflict with the Biblical account, literally read" and thus violated the Establishment clause.[15] From this point forward, the legal game changed. Now, the question was whether creation-based approaches constituted teaching religion in a way that violated the Establishment clause. Defending Genesis was off the table. Something else changed as well. Anti-evolution groups now carefully crafted approaches to education that aimed to evade the Establishment clause exclusion (and one sees this dynamic at work in the most recent critical inquiry laws). After *Epperson*, opponents of evolution proposed "balanced treatment": if evolution is taught, creationism must be taught for fairness. The U.S. Court of Appeals rejected balanced treatment in *Daniels v. Waters* (1975), arguing that by assigning a "preferential position for the Biblical version of creation" over "any account of the development of man based on scientific research and reasoning," the challenged statute officially promoted religion.[16] This Court of Appeals ruling was influential but not binding nationwide.

Until 1975, opponents of evolution successfully argued that evolution should be banned because evolution contradicted the creation story of Genesis, *or* creationism must be taught to counterbalance the effects of teaching evolution on particular religious beliefs. When *Daniels* ruled out both strategies, "creation-science" was offered as an alternative *scientific* theory. Creation-science claimed to supply scientifi c evidence for the biblical account of creation. In *McLean v. Arkansas Board of Education* (1982),[17] a U.S. District Court overturned Arkansas's Balanced Treatment for Creation-Science and Evolution-Science Act on the grounds that it also violated the Establishment clause. Five years later, in *Edwards v. Aguillard*, the Supreme Court struck down a similar law from Louisiana.

Like today's critical inquiry and academic freedom laws, the 1982 Louisiana Balanced Treatment for Creation-Science and Evolution-Science Act claimed to broaden science instruction and protect academic freedom. However, a six justice majority found that the statute lacked secular purpose and advanced a particular religious belief in violation of the Establishment clause.[18] Justice William Brennan wrote the majority opinion, joined by Justices Marshall, Blackmun, Powell, Stevens, O'Connor, and White. Justices Scalia and Rehnquist dissented. For the majority, creation-science depends upon a "supernatural being" and is, therefore, a "religious viewpoint." Brennan alludes to the historical relationship between "certain religious denominations" and anti-evolutionism but does not analyze the history in detail (as was done by the District court in *McLean* or the Supreme Court in *Epperson*). He explicitly holds that only certain religious "sects" or "groups" oppose evolution. Because Brennan has no *text* declaring Louisiana's intent to shore up religion, he refers to the historical link between fundamentalism and anti-evolution (noted earlier) in order to establish that the Legislature *intended* to favor religious establishment. For Brennan, Louisiana's mandate for creation-science advances "a religious doctrine by requiring either the *banishment of the theory of evolution* from public school classrooms or the *presentation of a religious viewpoint that rejects evolution in its entirety*." The Act's purpose is to discredit "evolution by counterbalancing its teaching at every turn with the teaching of creationism."[19]

Two crucial claims emerge from this brief discussion of the precedents used in *Dover*. First, dependence upon supernatural explanations defined creation-science as religion, facilitating exclusion from the science curriculum. Second, the Court relied on the history of evolution in America to determine intent. Because certain sects—past and present—view evolution as a threat to religion, courts must question the legislature's motivations. Louisiana's stated secular purpose in opposing evolution—enhancing student education—was (according to Brennan's opinion) a "sham." The Court excluded creation-science as religion but recognized that there are religious motives for discrediting evolution.[20]

In *Dover*, Judge Jones treats *Edwards* as a consensus decision, but Justice Antonin Scalia's dissent not only takes issue with the majority's claims regarding secular purpose and intent, it provides the roadmap for Intelligent Design as well as the new critical inquiry laws in Louisiana and Tennessee.[21]

Scalia argues that the legislative history and "historical and social context" demonstrate that the Louisiana legislature intended a secular purpose: "academic freedom" and the "search for truth." Scalia stridently frames creation-science as a remedy for censorship. Creation-science is "a collection of educationally valuable scientific data that has been censored from classrooms by an embarrassed scientifi c establishment." Teachers have been "brainwashed" by an entrenched scientific establishment that treats evolution like a "religion." It is illiberal—"Scopes-in-reverse"—to ignore creation-scientists' claims. Louisiana included creation-science based on the testimony of credentialed experts: "The witnesses repeatedly assured committee members that 'hundreds and hundreds' of highly respected, internationally renowned scientists believed in creation-science and would support their testimony."[22] Scalia concludes that the "body of scientific evidence supporting creation-science is as strong as that supporting evolution. In fact, it may be stronger." Because it is credible that the legislature believed teaching creation-science would increase student knowledge and encourage critical discussion, Louisiana's stated goal—to enhance education—must be taken at face value.

Scalia explicitly rejects consideration of the history of Christian fundamentalism and evolution. This is ironic since Scalia's dissent employs a tactic from turn of the century Fundamentalism. Compare Scalia with "Occupant of the Pew"—written in 1911 and reprinted in *The Fundamentals*. "Occupant" charges that Darwinism is not just dying but "already dead" and provides a list of scientists, "distinguished in their respective departments," that offer "unequivocal testimony" that the facts do not support the "theory" of evolution.[23] This type of attack, using discredited scientists to question the science of evolution, was used by the Fundamentalist movement for decades—and Scalia repeats it verbatim.

It is impossible to overstate the extent to which Scalia's dissent is a work of fiction. Creation-science does *not* have scientific evidence to support its claims. While it is true that a handful of renegade PhDs have published books, no creation-scientists have been able to publish in established peer-reviewed journals or provide scientific arguments for their claims. Only 0.15 percent of American earth or life scientists accept creationism.[24] Scalia ignores the remarkable *amicus* brief submitted by 72 Nobel laureates, 17 state academies of science, and 7 national scientific organizations demonstrating that there is no evidence—no *science*—supporting creation-science.[25] Scalia assumes that the testimony of a few PhDs is the same as scientific evidence that has been vetted by scientific journals or replicated in countless scientific studies for more than a century: giving *any* position a hearing defines his "search for truth." His claim that the body of evidence for creation-science may be stronger than evolution is patently absurd. If we follow Scalia's paradox of neutrality, schools would have to give "balanced treatment" to Holocaust deniers or race supremacists as long as they had PhDs.

Yet Scalia's dissent is important because his arguments mirror those made 18 years later by the ID movement in *Dover* as well as the new critical inquiry laws in Louisiana and Tennessee. In an ongoing reciprocity between civil society and courts, the creationists, the ID movement, and science organizations have all reacted to court precedents by tailoring their arguments and literature to respond to court tests and precedents. Scalia provided a roadmap for ID: frame the discussion in terms of free speech and openness (the language of democratic deliberation and rights); speak the language of science, find "experts" who will endorse creationism with no evidence; and never—ever—mention religion. The new critical inquiry laws read like a second edition of Scalia's dissent.

The *Edwards* case—and Scalia's missive—also woke up supporters of evolution. The national science community—particularly the National Academy of Sciences—began a careful and elaborate campaign to discredit creation-science, restate the evidence for evolution, insist on the compatibility of evolution and belief in God, and actively defend the teaching of best available science in public school science curriculums.[26] Since *Edwards*, the National Academy of Sciences (NAS) has devoted extensive resources to understanding how they can change elite and public opinion to support their message: extensive evidence supports the theory of evolution,

evolution should be taught in schools, and evolution does not threaten religious beliefs.[27]

UNPACKING DOVER: EXCLUDING RELIGION
AND DEFINING SCIENCE

In *Kitzmiller v. Dover*, Judge Jones references all of these cases but depends most heavily on *Edwards* as the most recent Supreme Court precedent. In order to exclude Intelligent Design as religion, he adopts two tests for the establishment of religion: endorsement and the religious purpose and effects prongs of the *Lemon* test.[28]

Endorsement, adopted by the Court after *Edwards*, asks whether a government's purpose is to endorse religion and whether the statute actually conveys a message of endorsement.[29] Justice Sandra Day O'Connor, the test's creator, is particularly sensitive to situations in which personal religious beliefs affect a person's standing in the *political* community: "Endorsement sends a message to nonadherents that they are outsiders, not full members of the political community, and an accompanying message to adherents that they are insiders, favored members of the political community." If a "reasonable observer" acquainted with the language, origins, legislative history, and implementation of the statute would perceive government to be endorsing or disapproving religion, the government action or statute is invalid. In *Dover*, Jones makes the case for endorsement by noting that opposition to evolution "grew out of a religious tradition, Christian Fundamentalism that began as part of evangelical Protestantism's response to, among other things, Charles Darwin's exposition of the theory of evolution as a scientific explanation for the diversity of species." He documents the "long history of Fundamentalism's attack on the scientific theory of evolution, as well as the statute's legislative history and historical context."[30] Given the historical context, the *Dover* court concluded that an objective adult, student, or *Dover* citizen would see ID as masked creationism: the Dover statement constitutes a religious position endorsed by the school.[31] Although Jones legitimately cites *McLean* and *Epperson* on this point, his claim that Brennan's conclusion in *Edwards* rests on a "thorough analysis of the history of fundamentalist attacks against evolution" is false. Making vague references to particular religious sects, Brennan's majority opinion avoided the language of *Epperson* or *McLean* in favor of a general history that did *not* highlight Christian Fundamentalism.[32]

Although Jones emphasizes endorsement (as a more recent Supreme Court test), he also briefly invokes the *Lemon* Test (relied upon by Brennan in *Edwards*). The *Lemon* test asks if a government action (1) has secular legislative purpose, (2) has as its primary effect the advancement of inhibition of religion, and (3) results in an "excessive government entanglement" with religion.[33] Unlike "creation-science," ID does not present scientific evidence in favor of the "Noachian Deluge" or refer to canonical Judeo-Christian narratives. Jones documents how ID texts were purposely stripped of references to any particular biblical text or tradition and concludes that the ID movement, directly responding to *Edwards*, strategically designed a creationism that could pass as secular.[34] On the basis of this evidence, Jones insists that the school district's motive was not secular and that the primary effect of the statement was advancing religion by challenging the legitimacy of evolution and providing an alternative approach.

* * *

Thus far, Jones follows the Supreme Court in relying upon the Establishment clause, though he updates *Edwards* by using O'Connor's endorsement test. Jones moves outside of Supreme Court precedent to make a second—very different—legal argument: Dover must exclude Intelligent Design because it is not science. In order to understand why Jones moves outside of Supreme Court precedent, we need to explore two theoretical nuances of ID.

Creation-science and Intelligent Design each contain two types of arguments: (1) criticisms of evolution as insufficient in explaining the origins of human life (the negative thesis) and (2) evidence of an alternative causal agent or Intelligent Designer of humans (the positive thesis).[35] In both *Edwards* and *Dover*, supporters of creation-science and Intelligent Design argued that teaching the weaknesses of evolutionary theory (negative thesis) *and* advancing alternative theories (positive thesis) enhanced student learning by broadening knowledge.

If creation-science and Intelligent Design are religion "masquerading" as diversity, courts can exclude them two different ways. The first approach, which I call *Defining Religion*, focuses on the positive thesis. The schools exclude as religion any teaching that relies upon a supernatural causal agent to explain natural phenomena. Brennan emphasized the positive thesis in *Edwards*. The second approach, *Defining Science*, highlights the negative thesis. The schools should include *only* material that is science. All other material—religious or not—should be excluded from a science classroom. In *McLean,* the District Court developed this *Defining Science* approach by identifying five essential characteristics of science: (1) It is guided by natural law; (2) It has to be explanatory by reference to natural law; (3) It is testable against the empirical world; (4) Its conclusions are tentative, that is, they are not necessarily the final word; and (5) It is falsifiable.[36] Based on the five criteria, the *McLean* court excluded creation-science because it was not science. Justice Brennan could have used either *McLean*'s five-point science criteria or arguments made by the *amicus* brief submitted in *Edwards* by 72 Nobel laureates in science, but he did not.[37] Instead, Brennan simply *Defined Religion* by rejecting creation-science as grounded in the supernatural.

In his *Edwards* dissent, Scalia insisted that creation-science provides scientific criticism of the theory of evolution. Relying on the word of a few testifying experts, Scalia never defined science or assessed whether creation-science met the criteria. Increasingly, Intelligent Design advocates insist that the current definition of science is the work of a prejudiced scientific establishment. Narrow and exclusive, the definition of science needs to be changed to include supernatural causation. In *Dover,* Jones brought in a host of expert witnesses to distinguish between science and pseudo-science. He returned to the argument from *McLean*: the criticisms of evolution presented by ID are "not science." Although Jones frequently relies upon *McLean*'s five criteria, he also outlines criteria specific to ID: a scientific theory can neither invoke supernatural causation nor claim that the inability to explain natural phenomena is grounds for supernatural causation. Additionally, a scientific theory must have academic credibility. If ID's attacks on evolution have been refuted by the science community and there are no peer review publications, testing, or research to demonstrate its validity, ID's critique of evolution is not scientific.[38] Thus, Jones excludes ID as "not science," using a combination of his own and the *McLean* criteria.[39]

Jones defends gathering information to prove that Intelligent Design is "not science" two ways. First, he sees his court as uniquely positioned to collect and interpret evidence from a wide variety of scientists, philosophers, historians, and so forth. Given the length and detail of the *Dover* trial, "the Court is confident that no other tribunal in the United States is in a better position than are we to traipse into this controversial area." More importantly, he claims that "we will offer our conclusion on whether ID is science not just because *it is essential to our holding that an Establishment Clause violation has occurred in this case,* but also in the hope that it may prevent the obvious waste of judicial and other resources which would be occasioned by a subsequent trial involving the precise question which is before us."[40] Quite disturbed by how ID has managed to become science, Jones appears keen to have *one* trial in which all the information is gathered so that, once and for all, ID can be dismissed from science classrooms because it fails the 5+ criteria. But his claim regarding establishment—ID must be "not science" to exclude ID as an establishment of religion—is false. Just because ID is not science does *not* make it religion.

I don't agree with Jones that finding creation-science to be "not science" is "essential" to the holdings in *Epperson* or *Edwards*. Brennan did distinguish promoting a theory that is religious from discrediting evolution. He acknowledged that, historically, certain religious sects had suppressed the teaching of evolution, and he implied that religious literalists oppose the teaching of evolution because they believe that it threatens their form of religious belief. But Brennan never addressed whether creation-science was science. He merely insisted that it was religion. Jones is up to something else. Anti-evolutionists believe that teaching evolution is a religious position: rejection of biblical literalism is tantamount to state-sponsored atheism. Teaching evolution threatens the belief in literalist religion because as more people believe in evolution, fewer will believe in biblical literalism. Jones inverts this fundamentalist literalist claim. If literalists believe that weakening evolution lessens the threat to religious belief, criticizing evolution (negative thesis) endorses religion. This is interesting (and clever), but it is not necessary to succumb to false logic (criticizing evolution implies a religious motive) or twist the meaning of *Edwards* to make an important point. Jones is right: pursuing the distinction—not science—has great legal and political potential.[41] First, it removes the need to establish a religious *motive* for passing legislation to include criticism of evolution in classrooms. Second, it pulls the discussion away from religion and moves it toward science. If Intelligent Design doesn't meet the criteria, it is out.

Yet Jones pragmatically relies on defining religion because *Edwards* is the controlling precedent. Here, he improves upon Brennan's analysis by clarifying the historical content. If fundamentalist Christian groups in the United States have historically identified evolution as a threat and organized to fight it, how should courts regard efforts to mandate the teaching of weaknesses of evolutionary theory? Jones considers the origins of fundamentalism to demonstrate the link. Jones wants courts to understand criticism of evolution in this particular historical context so they can forcefully challenge legislative motives. The motive for requiring the ID statement in *Dover* was religious whether God was mentioned or not.[42]

Jones wants courts to understand the intellectual history of anti-evolution. Fundamentalists claim that (1) belief in evolution is incompatible with belief in God, thus, evolution leads to atheism; and (2) if evolution is false, creationism is true.[43] Jones assumes that these claims prove non-secular purpose: ID criticizes evolution to protect religion rather than to contribute to scientific inquiry. In the case of ID, the positive and negative theses are *both* religious positions, and the positive thesis is implicit in the negative. Paralleling the two Fundamentalist claims, the negative thesis (1) weakens evolution and lessens the threat to religion and (2) creates an argument for divine or supernatural creation. Because anti-evolutionists use both negative and positive arguments to defend religion, courts should examine the beliefs of literalists in order to judge whether the intent of the laws is secular (improve scientific inquiry) or religious (protect literalist religion by criticizing evolution). Jones emphasizes the *intent* to inject religion rather than the school's responsibility to exclude bad science.[44]

In *Dover*, it was quite simple to prove intent. The district leadership had publicly called for bringing creationism back into the classroom and sought out ID to provide a religious alternative to evolution. The public debate was focused on religion in the schools, and one of the board meetings ended with a call to be born again. Astutely, Jones anticipates future cases in which *only* the negative thesis is presented in the legislative history and the text of the statute. If criticism of evolution becomes the *only* goal in a future case,

Defining Religion fails to exclude ID. The new critical inquiry laws present precisely this type of case as the statutes emphasize the negative thesis and list other scientific controversies in addition to evolution.[45]

Defining Science promises a solution. Even if no positive claims are made regarding a supernatural creator and there is no legislative history regarding intent, courts can exclude material based on a secular standard for valid scientific criticism. Categorizing ID's criticisms of evolution as "not science" weakens the claim that ID enhances student science education, and, more importantly for Jones, it establishes ID as a religious position disguised as science.

Defining Science can establish criteria for scientific data, but it takes two very different forms in *Dover*: (1) anything that is "not science" cannot be taught as science, and (2) anything that is "not science" is religion and can be excluded using the Establishment clause. The former avoids the defensive strategy and fraught judgments of motive, but Jones presents no support for this version of *Defining Science* in Supreme Court jurisprudence or the Constitution. Though he emphasizes the *McLean* criteria, *McLean* (as a District Court case) is not Supreme Court precedent, and Jones appears to doubt the ability of *Defining Science* to stand alone. He occasionally accepts a false claim of ID—science and religion are incompatible—so that he can maintain that if ID is "not science," it is religion.[46] Here, he slips into a fallacious argument made by the ID movement, and *Defining Science* becomes another way to defend against religion. Jones is right that courts need to define science, but he does not supply a constitutional argument.

This analysis of *Dover* yields two important insights. First, *Dover* addresses both the positive and negative theses of Intelligent Design using categorical determinations to exclude ID. *Defining Religion* claims that dependence on a supernatural creator defines religion. Thus, teaching evolution can be *defended* by excluding positive religious claims. *Defining Science* provides a powerful—and appropriate—tool that does not require a discussion of religion, but *Dover* failed to show how the Supreme Court might support the approach. In its suspect form, *Defining Science* demonstrates that a claim lacks scientific credibility then *infers* that the intent to include it was religious. Once religious intent is established, ID can be excluded as an establishment of religion. The courts need to find an appropriate way to Define Science, but this suspect approach should be rejected.

Second, *Dover* never makes a positive case for including science. Neither *Dover* nor *Edwards* suggests that the state or locality should require the teaching of best available science in public schools for political citizenship: reject ID for the public good. Only the *amicus* brief from the Nobel laureates and science organizations offered a vision of science, politics, and education. Focusing on national defense, food production, critical thinking skills, and problem solving for public health and the public good, scientists and science organizations articulated *positive* reasons for teaching the best available science—the theory of evolution—in public schools. In this *amicus* we see the beginning of an approach that defines science education as crucial to citizenship and the public good in a democracy rather than defending science against religion.[47] Before exploring how the Supreme Court might adopt such an approach (Chapter 3), we turn to the ways that democratic theorists treat science, evolution, and education.

2 Battling Creationism with Deliberation, Autonomy, and Liberal Values

As states move to pass critical inquiry statutes—rather than requiring the teaching of creation-science or Intelligent Design, courts will find it harder to exclude discredited science from classrooms using the Establishment clause. In the last chapter, I suggested that liberal democracy needs an alternative legal approach. This chapter uses three contemporary liberal justifications of democratic education—based on deliberation, autonomy, and liberal values—to explore what is at stake for liberalism when science education is questioned by Intelligent Design or critical inquiry. Even as Amy Gutmann, Harry Brighouse, and Stephen Macedo provide valuable insights, none of these theorists provide a principled *and* politically viable argument for including evolution education.

FREE EXERCISE OF RELIGION: THE RIGHTS OF PARENTS AND CHILDREN

All forms of liberalism emphasize the freedom of the individual to craft a life and balance individual liberty against the rights of others. Liberals diverge over the extent to which the choices of individuals must be enhanced, supported, or regulated by the state. At one extreme, individuals should be free to pursue whatever they wish, including self-harm or ignorance (libertarianism). At the other, differences in resources, gender, race, and natural talents oblige the state to supply an education necessary for genuine choice. At its most extreme, liberalism asks the state to shape individuals to choose autonomy as a value for private life (comprehensive liberalism).

Liberals struggle to balance supporting meaningful choice through education with their fear of state authority. In modern societies, most students come to school with powerful identities based on the ethnic, religious, cultural, and linguistic identities and practices of their families and communities. Diversity liberals, such as William Galston, assume that public education must respect these collective identities—deferring to the wishes of parents—and refrain from teaching material or requiring education that threatens religious belief or practice. Political liberals, such as Stephen

Macedo and Brian Barry, object, believing that diversity liberalism fails to sufficiently differentiate the interests of the child from the family or birth community. Correctly understood, pluralism supports a genuine choice made by the student who will become an adult rather than the preferences of parents, families, or communities. If the individual matters, the state must supply an education that allows the child to critically evaluate the life they have inherited, including the ability to exit their birth community.[1]

Liberal decisions regarding education depend upon how children become reasoning and choosing adults. If children spring from the earth "like mushrooms fully loaded" with the capacity to choose and give consent, the state can leave them alone. But liberals like John Locke and John Stuart Mill believe that children—born *to* reason and equality—need training and education to become capable choosers of their own destiny.[2] Locke and Mill assume that, generally, families will act in the best interests of their children by educating them, but Mill cautions that parents often act against the best interests of their children. The child needs "an education fitting him to perform his part well in life toward others and towards himself," and the state "should require and compel the education, up to a certain standard, of every human being who is born its citizen." Even though Mill fears the state educating all children, he believes it is a "moral crime" to leave the child uneducated because true liberty depends upon education.[3]

American courts use "compelling interest" to capture this tension between the state, parents, and children. When the state requires children to act in ways that their parents object to, the state must demonstrate either an important *collective* interest that can override the free exercise rights of the parents, or the state must claim to protect the *individual* interest of the child. American courts—like liberal theorists—never grant parents absolute discretion over their children. Courts, for example, provide medical help to children even when parents claim that their religious exercise forbids a blood transfusion. As a guardian of the child, the liberal state balances the rights of parents and children when they conflict.[4] Liberals struggle to define *when* children have rights to vote, drink, stop attending school, drive, or override their parents' decisions.

Education poses thorny questions for liberal theorists. Well aware that states can use schools to indoctrinate, solidify power, assimilate immigrants, establish religion, or encourage particular forms of citizenship, liberals must articulate the goals—both collective and individual—of public education. In the case of science education, American schools face powerful resistance from parents who believe that evolution assimilates their children to secularism by contradicting a key assumption of their faith: the literal nature of the Genesis narrative. Parents and, increasingly, legislators demand that schools caution students about the truth of evolution or provide discussions of alternatives to evolution that are more acceptable to their literalist faiths.

AMY GUTMANN'S DEMOCRATIC DELIBERATION: CREATIONISM AS OPPRESSION

Amy Gutmann, now president of the University of Pennsylvania, astutely insists that teaching creationism negatively affects democratic practice and citizenship. Ultimately, Gutmann fails to provide a formula that can exclude Intelligent Design or new Academic Freedom laws requiring "balanced" treatment of scientific controversy. She reproduces the limitations of the logic of the *Edwards*'s decision by *defensively* excluding religiously based science as oppressive and fails to supply positive criteria for including the theory of evolution—or any science—in the school curriculum.

Although Amy Gutmann's work predates the Intelligent Design movement, her valuable analysis of creationism in the context of democratic citizenship helps clarify why Intelligent Design now claims to be a minority view with the "right" to be represented in public school curriculum. Her insights are particularly valuable with regards to democratic input in education decisions and the limits on democratic decision-making in a pluralistic society.

Gutmann defines *creationism* as a religious creed that threatens the ability of students to deliberate as *equal* citizens. Her approach mirrors Justice O'Connor's jurisprudence of endorsement. For O'Connor, endorsing religion in the public school context creates insiders and outsiders in the political community.[5] Gutmann fleshes out the implications of endorsement, recognizing that equal citizenship requires public spaces in which participants are not favored based on religious belief or non-belief. If creationism—a belief held by a literalist Christians—is taught as fact in a public school science classroom, citizens do not deliberate equally. Gutmann characterizes creationism as a form of oppression that may be excluded based on her desire to protect democratic deliberation in the liberal state. Because deliberation is the core of democratic citizenship, democratic education should "cultivate the skills and virtues of deliberation."[6]

Not only should the *substance* of education be directed at the needs of a democratic polity, the *process* of making decisions should be democratic as well. Majorities should have a substantial role in making educational decisions for the training of democratic citizens in their communities. Gutmann favors decisions that are local and democratic as a condition of being fully deliberative.

Yet Gutmann is acutely aware of the perils of majority rule: the danger that majorities (e.g., religious, intellectual, or ethnic) can intrude on the rights of individuals. Despite her support for democratic decisions, Gutmann never presumes that democratic deliberation will always produce desirable results. Majorities can be "undemocratic" because educational policies can directly or indirectly "stifle the capacity and even the desire for deliberation" in individuals. For example, some states try to protect "contestable political perspectives" from intellectual challenge. Given the potential for oppression, Gutmann restricts the local majority's educational authority if it restricts "rational deliberation on contestable perspectives or excludes potentially educable citizens from an adequate education." Social reproduction is a legitimate—and crucial—element of democracy, but the liberal democratic commitment to "nonrepression" and "nondiscrimination" prevents "states, and any group within them, from restricting rational deliberation of competing conceptions of the good life and good society."[7] Here, Gutmann grapples with democratic and liberal elements of American political practice.

Gutmann would allow a state or school district to teach American history to build common democratic values—even history that is overly generous to the American side. But she would prevent a local school district from introducing a racist or segregationist ideology through pedagogy, textbooks, or curricula. She returns to endorsement and equal citizenship. When the public school classroom endorses racism, the student of color is no longer an equal citizen: her potential for rational deliberation is destroyed—as well as the public discussion of the wider group. In the case of racism or overt Christian teachings, Gutmann's approach is intuitively compelling, yet her terms—contestable political perspectives, rational deliberation, and adequate education—are not well-defined enough to settle more controversial cases such as Intelligent Design.

Proponents of Intelligent Design, as noted earlier, exploit the language of democratic deliberation to claim that ID is an oppressed minority voice seeking to enrich discussion and deliberation by providing an alternative scientific approach to the theory of human origins. At first glance, Gutmann's criteria for democratic education—especially her support for democratic process and the inclusion of competing conceptions—seems to *support* teaching of Intelligent Design in public schools. In her discussion of textbook selection, Gutmann requires openness to citizen participation, especially the "potential to open citizens to the merits of unpopular points of view."[8] In the *Dover* case, the process for including ID was local, democratic, and deliberative. Citizens elected the school board that created the ID policy. Citizens could, and did, attend open meetings to deliberate before the policy was enacted. Some board members used the language of democratic inclusion: challenging students to reflect upon a competing theory such as ID would enhance diversity and deliberation. In *Dover*, the process and outcome seems to support Gutmann's criteria: a democratically elected school board supplied supplemental resources on a controversial approach to human origins after holding public meetings.

But Gutmann would likely reject ID as enhancing deliberation on human origins based on her requirement that public education cannot oppress individual students. For example, she rejects the teaching of creationism as "indirect repression" because "the broader implications of the policy rather than the policy itself are repressive." According to Gutmann, a democratic decision to teach creationism can—and must—be rejected because

(1) curriculum should be selected by professionals not majorities, (2) religious pluralism requires the exclusion of teaching a Christian theory of creation, (3) there exists an American consensus to accept secular standards of reason, and (4) the scientific case against creationism is straightforward.[9] Although thought-provoking, Gutmann's four criteria pose problems for excluding Intelligent Design.

In terms of curriculum content, Gutmann never clarifies *who* can make a "professional" decision: teachers, school board, a professional organization within the discipline, or the nation, state, or local authorities? She seems to gesture toward depending on professionals within a field as she distinguishes between state-of-the-art information and religious indoctrination, but she complicates the case when she insists that a public education curriculum is *not* based on solid evidence and reason alone. If we depend upon the professional standards of history, public schools would be forced to offer a more balanced treatment of the American Revolution—yet Gutmann allows the more nationalistic account. She distinguishes creationism from the American Revolution on the grounds that creationism is "believable only on the basis of a sectarian religious faith." If we teach creationism as science, we indirectly violate the principle of nonrepression because we impose "a sectarian religious view on all children in the guise of science."[10] If we teach the American Revolution in a biased way, we deprive students of full knowledge of events and reinforce national myths, but we do not impose a religious position. Gutmann admits that a state or local school district may mandate history that is not "best"—but the same majority cannot impose a particular religious view. Here, Gutmann's analysis rests on excluding creationism as religion: the Establishment clause approach found in *Edwards*.

Gutmann's second criteria is pluralism. Creationism cannot be taught in a pluralistic religious context without oppressing some individuals and depriving them of their equal status as deliberating citizens. But, in a hypothetical society with homogenous beliefs, "teaching creationism might be compatible with *their* democratic standards."[11] Here, Gutmann side-steps two important problems. First, there is no such thing as a religiously homogenous society; all religious groups have important areas of departure and disagreement. Even if there was a community that was generally homogenous, it would be essential—if we believe deliberation is essential to democratic citizenship—to teach a variety of views rather than solidifying the majority's understanding. Second, creationism is an easy case because most creationists openly name Christian religion or Genesis as the source of their belief. Intelligent Design carefully avoids attachment to any particular religion's view of human origins. Even if ID's intent is religious, ID seeks credentialed scientists: their claim is that they represent a scientific approach.[12] We end up with two arguments for pluralism: religious pluralism denies creationism access to public schools, yet "scientific" pluralism might include Intelligent Design.

Gutmann may have a way out—because she limits the contestable perspectives that must be given voice to the "political." If this is true, teachers could discuss the creationist movement or ID in a social studies class in the context of American history or political culture. Yet schools could not present creationism or ID in a biology class as an alternative to evolution. Here, Gutmann's argument is stronger. Engaging the controversy over evolution, creationism, and Intelligent Design in the context of American history and politics might well enhance political deliberation. Yet Gutmann might leave room to reject the utility of presenting creationism and ID as "contestable perspectives" in a biology lecture.

Gutmann does not push on this distinction because she relies on excluding creationism rather than justifying best available science as a part of democratic education. First, she is not responding to ID or critical inquiry statutes. Second, she is confident that science curriculum will be determined by "secular standards of reasoning." Creationism can be rejected because the United States has "widespread acceptance" that "scientific standards of evidence and verification" should guide the content of knowledge transmitted to future generations. According to Gutmann, religious pluralism does not interfere with the scientific standards because "[m]ost Americans have reconciled the tenets of their faith with the findings of science." Given her confidence that Americans accept scientific evidence as the criteria for judging curricula, Gutmann can claim that the "scientific case against creationism is straightforward."[13]

Excluding creationism or ID is more complicated because, in the case of evolution, Americans neither accept science as the standard nor the clarity of the case against creationism. A brief overview of polling evidence helps clarify why Gutmann's claims for democracy and a democratic consensus are at odds.

Over the last 30 years, public opinion polls—using multiple formats and wording—establish that 45 percent of Americans accept the theory of evolution, while 46 percent support creationism: God created humans fully formed. Those who believe in creationism show more far more certainty: 30 percent *definitely* believe in creationism, while 16 percent believe it is *probably* true, whereas only 15 percent of Americans believe that the theory of evolution is *definitively true,* while 30 percent believe it is *probably* true.[14]

Americans also split over whether evolution is well supported by scientific evidence (34%–35%) or just one theory among many (35%–39%). If Darwin is mentioned, belief in a scientific consensus regarding evolution goes down.[15] When asked whether fossil evidence proves that the theory of evolution is true, the numbers remain roughly the same, and, again, those who reject the evidence are more confident than those who accept it.[16] Despite all the data supporting the theory of evolution, only 35 percent of Americans think that evolution is well-supported by evidence. While the National Academy of Sciences insists there is an "overwhelming" scientific consensus on the theory of biological evolution, *only half of Americans believe that scientists have such a consensus.*[17]

Support for evolution is strongest among men, whites and Latinos, people under 30 and 30–49, college graduates or those with post-graduate training, Easterners and Westerners, and those who claim "education" was the biggest influence on their views about how life developed (as opposed to personal experience or religious belief).[18] Support for evolution is highest among Jews, Catholics, white mainline Protestants, and the religiously unaffiliated and lowest among white Evangelical Christians and black Protestants.[19]

How do these polling statistics help us evaluate Gutmann's claims? Gutmann assumes that there is an American consensus: scientific standards should be used to evaluate curriculum. Among scientists and people with post-graduate degrees, Gutmann is correct. Yet Gutmann overstates national confidence in science. Roughly split, Americans have "widespread acceptance" of *both* science and religion as a standard as the American *public* does not believe either that there is sufficient data to prove the theory of evolution or that a scientific consensus exists. A supermajority—about two-thirds of Americans—want schools to teach *both* creationism and evolution. Only 15 percent confidently support teaching evolution alone, and 30 percent want creationism *instead* of evolution.[20] Americans are outliers compared to other liberal democracies where clear majorities believe in evolution and only small minorities reject evolution as false.[21] Only half of Americans accept Gutmann's claim that the case against creationism is "clear."

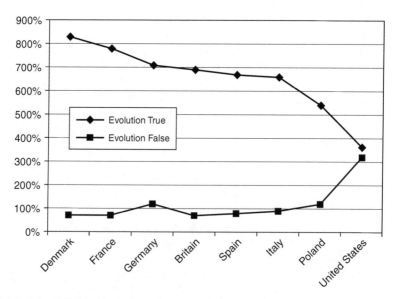

Figure 2.1   Belief in Evolution: Europe and the United States

Amy Gutmann connects education, the reproduction of democracy from generation to generation, and the equality of citizens as she justifies education as the foundation of public reason. Democracies should cultivate "common, secular standards of reasoning among citizens." Although such standards are not neutral to all religious beliefs, they are "a *better* basis upon which to build a common education for citizenship than any set of sectarian beliefs" because they are a "fairer and firmer basis for peacefully reconciling our differences." Democracies are not "bound to teach the truth," but they are required *not* "to teach false doctrines that threaten to undermine the future prospects of a common democratic education."[22] Gutmann is correct: democracy cannot tolerate teaching false information to citizens. She rightly emphasizes the importance of deliberation and the need for reproducing democratic principles and practices through education, but she does not explicitly discuss how accurate science education contributes to public reason or how teaching discredited science harms public reason.

Moreover, Gutmann does not consistently argue her case in terms of the necessity for public reason. In a diverse society, Rawls insists that citizens must justify their *political* positions and votes using public reasons that are neutral with respect to comprehensive moral, religious, and philosophical doctrines.[23] Gutmann claims that the *scientific* case against creationism is straightforward, but the *democratic* case is complex: "we cannot conclude that, having determined that the evidence favors evolutionary theory over creationism we will know what belongs in the classroom and what does not."[24] If Gutmann seeks to protect the deliberative discourse that supports liberal government, why not reject creationism on the grounds that it is false? Why not frame these challenges to science education in terms of protecting the reason necessary for public discourse?[25]

Gutmann wants students (as future citizens) to think critically and collectively about politics. Aware that it is tricky to write a curriculum to cultivate such skills, she insists that teaching creationism thwarts the promotion of common, secular standards of reasoning and then discredits public schools "in the eyes of citizens whose religious beliefs are not reflected in the established curriculum."[26] Here, Gutmann relies on the defensive strategy—used by the Supreme Court in *Edwards*—to exclude creationism as religion that oppresses. Because she relies on defending against creationism, she fails to supply a positive criteria for including the theory of evolution—or any science—in the school curriculum. She demonstrates how religion damages democratic deliberation, but she does not address how science furthers deliberation or political reproduction.

In part, Gutmann does not see the need to defend science because she relies on a—non-existent—consensus regarding secular standards of reasoning. The new critical inquiry laws—using Justice Scalia's logic from *Edwards*—capitalize on the lack of consensus. These laws present Intelligent Design or creationism as alternatives to evolution that open up discussion and improve student learning. Like Scalia, they reject the neutrality of Gutmann's criteria—shared intellectual standards of reason—and claim that evolution is anti-religious.

But if there is no such consensus in the United States, how might democratic education—in the name of supporting democratic reason—teach students to accept "shared intellectual standards" for science education? Gutmann astutely connects controversies over teaching creationism with fundamental issues of democratic citizenship, but she ultimately fails to provide a formula that can exclude Intelligent Design or new Academic Freedom laws requiring "balanced" treatment of scientific controversy.

AUTONOMY FOR THE FLOURISHING INDIVIDUAL

Whereas Gutmann connects education with the maintenance of democratic politics, Harry Brighouse believes education should support the individual: the student. Brighouse brilliantly specifies the *skills* necessary for an individual student to flourish, but his education for autonomy approach cannot find consensus among American liberals—or American courts.

For Brighouse, students should not simply inherit the choices of their parents, learn a little at school, and return to their communities. In order to guarantee genuine and meaningful pluralism, liberal education must train students to choose the type of life that they will lead—one that is suited to that individual's personality. To flourish, humans must be able to evaluate and challenge the religious and cultural views of parents, be self-supporting participants in the economy, and be effective, reasoning participants in public decision-making. As such, students have a "right" to learn about *different* ways to live so that they may reflect on their own way of life in light of those alternatives. Schools serve as a public resource for self-construction, and they must help students learn to deliberate critically. Moreover, education should empower students to *act* on those reflections and, if they chose, "revise or reject the way of life their parents would pass down to them." Thus, it is not sufficient to grant adults the freedom to worship and ask the state to respect these identities as Galston suggests.[27] Meaningful pluralism and the authenticity of private life rely on the public sphere providing a resource for the construction of the self, including the ability to exit.

As individuals evaluate their way of life, they must engage with religion, culture, and work—as well as politics. For Brighouse, there are fruitful spillovers between moral decision-making and political citizenship because "political reflection cannot be neatly differentiated from the skills involved in evaluating one's own way of life."[28] Political participation requires a spirit of respect and a willingness to engage in public reasoning.[29] Although he underlines the effects on the individual, if education succeeds in facilitating individuals' capacity to flourish, this affects the political lives of others: "We exercise our powers of self-control, of rational thought, and of altruistic concern by being good citizens, and we also earn the respect of others; these things are genuinely valuable to us as well as to them."[30] Individuals' actions affect the lives of others.

Brighouse welcomes political participation, but he speaks cautiously about work. Although economic self-sufficiency is essential to most forms of human flourishing, economic competence is often treated as a means to a collective end: national economic success. Brighouse highlights the role economic competence plays in individual flourishing while rejecting economic *growth* as a legitimate democratic goal. Thus, economic self-sufficiency should be the means—not the end—of a rewarding life. Humans should never be considered human capital.[31]

What kind of education is necessary for this ability to reflect and govern oneself rather than be ruled by past traditions or family decisions? Brighouse emphasizes the ability to *evaluate* as crucial to "uncovering how to live well." In particular, students must learn to identify fallacious arguments of all types and to distinguish between fallacious and non-fallacious arguments. The autonomous person needs to be able to distinguish between appeals to authority and appeals to evidence and between inductive and deductive arguments, as well as to identify *ad hominem* arguments and other misleading rhetorical devices." Brighouse does not emphasize science education, though he acknowledges that math and science are "great cultural achievements" that should be included in the curriculum because they assist individuals in making decisions related to leading a "good life."[32]

Brighouse's autonomy-based approach has had limited traction in American courts, but a famous dissent by Justice William O. Douglas embraces human flourishing and education for choice—emphasizing the child rather than collective interests. In *Yoder v. Wisconsin* (1972), three Amish families insisted that Wisconsin's compulsory education law infringed upon their free exercise of religion; the required two years of high school interfered with Amish spiritual rejection of worldly goods and commitment to simplicity in daily life. Chief Justice Burger's majority opinion emphasizes the public stake in educating students but ultimately agrees, based on free exercise claims made by the parents, that Amish teenagers are exempt from education beyond the eighth grade. Douglas, the lone dissenter, focuses on the children as individuals. If the Amish believe that ignorance of modern ideas makes children more likely to follow the religious beliefs of their parents, Douglas has little sympathy. Douglas affirms education for choice: the student should be "masters of their own destinies" who do not have their parents' "notions of religious duty" imposed upon them by force. Children should not be "harnessed" to the Amish way of life by parents who have "authority" over their children. Lack of education affects all aspects of the individual throughout the course of his life: "[I]f his education is truncated, his entire life may be stunted and deformed." Douglas pushes the Court to distinguish the interests of the *children* from the free exercise rights of their parents. A child who wants to be an astronaut or oceanographer needs to break with Amish tradition and go to high school. Douglas suggests that the state must ask each child if she or he wishes to go to high school and, if necessary, override the parents' desires. Thinking about the professional choices of the students, Douglas appeals to examples from the sciences. Douglas focused on the right of the child to be exposed to ideas, even those that challenge the religious traditions of their parents.[33]

The *Yoder* majority portrays the Amish as hard working and law abiding, but the Amish also refuse jury duty and demand exemptions from regulations regarding midwifery, Occupational Safety and Health Administration (OSHA) standards, and vaccinations. Douglas wants the child educated beyond the eighth grade to ensure that the child can reject—or embrace— the life that his or her parents chose for him or her. Given that the Amish require workers to opt out of social security (in favor of a community-based support system) or be shunned, Brian Barry labels them an involuntary association. The Amish *can* have an illiberal internal requirement, but Barry questions whether an eighth-grade education creates a realistic exit option for children and future adults.[34] Even a diversity liberal like Galston believes the state must protect the individual from being coerced into or trapped within a way of life. Such state protection includes education to be aware of life alternatives, analytically capable to assess the alternatives, and fit for participation in another form of life. The individual should also be free from brainwashing and psychological coercion.[35] Because Brighouse, Douglas, Barry, and Galston assume that the child should choose the Amish religion *freely*, they empower the state to provide education for individual choice (though they part company on what that education might look like).

The autonomy rich, child-centered approach offered by Brighouse and Douglas has never been supported by even a plurality of Supreme Court justices. Even many liberal theorists reject it as insufficiently sensitive to the deep divisions about the meaning of the good life that defines liberal pluralism. Thus, education for autonomy is often criticized as insensitive to the views of the good life that do not privilege or require evaluation and autonomy (e.g., Galston, Macedo). Critics of Brighouse and Douglas's approach reject the "good life as autonomy" as a sectarian view that cannot be imposed by the state.

The central problem with this approach is that many liberals do not accept comprehensive liberalism or autonomy as the central value of liberalism. The Supreme Court reflects this tension with few justices ever taking such an approach—though many do speak to both collective and individual concerns. But Brighouse provides an approach to liberal education that does not defend against religion—and one that leads us to think about the student as an individual rather than the rights of the parents. Brighouse carefully specifies the *skills* necessary for an individual to make choices—and requires schools to teach those skills. The frame—individual autonomy—is wrong, but the content—skills—seems a more fruitful approach given the challenges posed by science education. Theorists such as Galston see no difference between autonomy and critical self-reflection, but the work of Barry and Brighouse suggests a possible approach based on critical reflection rather than comprehensive autonomy.[36] As I will demonstrate at the end of the chapter, a skills approach may be joined with political citizenship to yield a stronger case for liberal education—including science education—that can be agreed upon by more theorists and judges.

STRONG VALUES: LIBERALISM WITH A SPINE

A third approach shuns autonomy in favor of defending the liberal state from illiberal groups who reject "core liberal values." For Stephen Macedo, political reproduction requires students to be exposed to "core civic aims" to the extent that the state may override the liberty of parents who wish to exclude their children from exposure to ideas that they believe threaten their private beliefs.[37] Macedo appears to rely upon Rawls's claim that liberty of conscience is limited by the "common interest in public order and security" as well as the state's need to deny toleration to the intolerant if the principle of toleration *itself* is threatened.[38]

Macedo dismisses liberalisms centered on diversity or difference that require the state to prioritize the desires of parents and communities to reproduce their views or tolerate illiberal ideas. Instead, Macedo favors a liberalism of assimilation: an "inescapable and legitimate object of liberal policy" that depends on the "justifiability of the values" that are being thrust upon groups that do not agree. In order to secure the "freedom of all," families cannot opt out of education that exposes their children to "very basic liberal virtues." Macedo recognizes that some individuals and groups will be disproportionately burdened by this requirement, but he accepts the effects because liberal assimilation secures the values that undergird the liberal state such that the fact of a burden on religious belief cannot trigger an automatic exemption from a public requirement. Liberalism should "grow a spine" and stand up for its core values, especially toleration.[39]

To illustrate how liberal values can be taught in spite of the religious objections of parents, Macedo explores a federal case: *Mozert v. Hawkins* (1987). When a Tennessee county mandated a literacy program that included a multicultural reader, some parents sought exemptions for their children because they believed that the readings threatened their born-again Christian beliefs.[40] Distinguishing exposure and indoctrination, a federal court unanimously ruled that schools *could* teach "ideas and concepts" that were "essential to democracy" as long as schools avoided "religious or anti-religious messages."[41]

For Macedo, *Mozert* demonstrates how liberal values—especially tolerance—can and must be taught.[42] But *Mozert* cannot support Macedo's political assimilation approach because it has a much more limited scope. Macedo fuses the meanings of autonomy, individuality, and critical thinking, and he routinely invokes values that are "shared," "very basic," or "core" without defining his concepts.[43] There is no clear consensus—in the courts or even among liberal theorists who support John Rawls—as to the liberal values that could be listed and, therefore, aggressively taught to children. As I will address later, liberalism relies on the capability to think critically and assess as a voter, juror, or soldier—as a potential co-author of the social contract. Macedo fails to distinguish imposing a value from teaching critical thinking skills.

Ironically, Macedo's strong liberalism proves weak when it comes to teaching evolution in the public schools. He narrowly defines the *content* of the education that would inculcate liberal values. When he dismisses art and science in favor of reading or the "knowledge of the diversity" that "constitutes our history and the importance of tolerance," he fails to appreciate the full range of disciplines—the liberal arts, including humanities and sciences—that contribute to the concepts and knowledge necessary for public reason in a democracy. Macedo—once—acknowledges a "critical attitude" as a core political "virtue."[44] But he fails to explore the extent to which science education that demonstrates the value of experimentation helps shape public reasoning skills. Macedo also ignores how public discussion depends upon *justifying* our positions: reasoned argument requires knowledge. Public education without enough information fails to give students robust choices: content and public reasoning are inextricably connected. When we put forward information that is false—as in the case of Intelligent Design or creation-science—we damage "the conditions of mutually intelligible public discourse."[45] As Macedo avoids a "holy war" between "religious zealots" and "proponents of science and public reason" by dismissing science as not "central" to core liberal values, his liberal values approach proves spineless when it comes to teaching science.[46]

Why does Macedo's stridency disappear when it comes to science? Like many liberals, Macedo fails to distinguish between avoiding positivism and embracing accurate scientific information for public discussion. For Macedo, science claims to provide a neutral standpoint by transcending subjectivity, and Macedo wants no part of this "John Deweyism, or what-have-you." Macedo refuses to endorse secular humanism or the scientific study of religion. Reasonable people, he claims, may believe that "in some areas science pulls up short," and liberalism cannot identify just "where or when." Given these concerns, Macedo refuses to associate science education with the maintenance of liberal values.[47]

As he runs from science as positivism, Macedo misjudges liberalism's ability to assess where science "pulls up short," and he inadvertently contributes to ID's manipulative use of scientific vocabulary.[48] We *can* distinguish scientific knowledge that is proven by data over a long period of time and those theories that are so contested as to be unsuitable for public school instruction. We can—through scientific method—demonstrate that Intelligent Design or creation-science is false. Macedo's concerns about positivism lead him to ignore the danger posed by fundamentalists who wish to inappropriately discredit good science and "enhance" education by providing "more knowledge." Indeed, his science "pulls up short" claim mirrors those made by the new critical inquiry laws in Louisiana and Tennessee: evolution is just a theory, so including other theories—even if they are discredited pseudo-science—improves the education of students by exposing them to "controversies" and "diverse" opinions. Unless liberals distinguish science as a neutral standpoint for solving social problems from science as a source of knowledge and critical thinking skills relevant to public discussions, they risk yielding the language of democracy to fundamentalists who look to censor public discussion in the name of democratic inquiry. Macedo's liberal education fails because he emphasizes liberal values at the expense of defining critical inquiry and democratic debate.

TOWARD A THEORY OF CRITICAL TEMPERAMENT

This analysis of liberal approaches to democratic education yields several important insights.

Liberalism cannot concede the language of democratic politics—of critical inquiry and open discussion—to Christian literalists. In light of new legislative trends, liberals must actively reclaim and clarify the relationship between education, public deliberation, critical inquiry, and science. Scientific method and knowledge affect the arguments that we make in public deliberations, and students must understand how to distinguish among claims. The "critical inquiry" laws in Louisiana and Tennessee insist that *all* theories are worthy of discussion in a science classroom. Liberals must overcome their fear of positivism and carefully delineate how public deliberation relies upon critical inquiry and knowledge. They must refuse to concede that all theories must be considered or all data is equally accurate in the context of public school education. We have meaningful ways to evaluate information, and public deliberation depends upon that knowledge.

Liberals are at their best when they resist arguments for liberal values or autonomy. Instead, they should advocate education to support public reason, liberal institutions, and democratic practices. Macedo is correct: liberalism should aggressively defend its foundational principles. But Macedo's value approach does not have sufficient support among liberals and, ultimately, cannot protect science education given Macedo's unreasonably thin conception of curricular content and his failure to focus on skills or critical inquiry. Many disciplines contribute to our capacity to question and argue—and science is one of them.

Brighouse's education for human flourishing and autonomy is compelling and complex, but the United States cannot educate individuals to choose autonomy as the governing principle of their value system. Neither the theoretical nor the political consensus for a liberalism of autonomy exists. Yet

Brighouse presents a rich and successful approach, focusing on the *skills* necessary for making choices. Genuine pluralism, he maintains, requires the capacity to make independent evaluations, including the decision to exit a birth community. Brighouse's emphasis on skills may be redirected to political citizenship in fruitful ways, which I will explore in Chapter 4.

Gutmann's attention to process and public deliberation holds promise, but she does not adequately engage with the role science plays in public reason. She excludes creationism as religion that oppresses rather than positively examine the role of scientific method and knowledge in public deliberation.

Liberals should respond to these new laws governing "controversial science" and "critical inquiry" by returning to the concept of public reason and insisting that education plays a crucial role in developing *both* the skills necessary for evaluating and deliberating as well as the knowledge necessary for making reasonable public arguments. Rather than focusing on autonomy, virtues, or values, liberal education should address how education affects the development of a democratic temperament, including the critical thinking—critical inquiry—capacities that nourish liberal citizenship, economic fitness, and ethical decision-making. Because citizens disagree about values and the good, liberalism must focus on the skills necessary to establish and maintain liberal democracy.

In the next three chapters, I develop my argument for liberal education for a democratic temperament. I begin with a set of arguments made by the Supreme Court (Chapter 4), and then I return to foundational American political thought (Chapter 5) to demonstrate how this understanding of democratic temperament—of the self-governing individual—is grounded in the American political tradition and how this approach might inform a contemporary reframing of critical inquiry.

3 Science and the Constitution
Creating an Education for Citizenship

As opponents of evolution shift their legislative strategy from attacking evolution education to advocating that students be exposed to diverse ideas, the need for an alternative jurisprudence that may be implemented by supporters of evolution becomes increasingly critical. This chapter links an older and established jurisprudence of education for citizenship with *amicus*—friend of the court—briefs by supporters of science education in order to create a jurisprudence that upholds the necessity of science education in the formation of citizens capable of maintaining and benefitting from a democracy.[1]

Throughout the last 75 years, the Supreme Court has repeatedly insisted that public education socializes American citizens and workers, reinforces liberal democratic norms and general constitutional principles, and undergirds pluralism. The Court has primarily focused on collective interests: the *nation's* need to shape individuals to provide military service, obey the law, and articulately criticize the government when it acts tyrannically. Yet the Court also recognizes benefits for the individual's financial advancement, ability to exercise rights, and generally flourish as a person. Mindful of both the collective and individual benefits of education, I use these precedents to craft a general theory of liberal education as the foundation for the political, economic, and private life of the individual and the reproduction of the liberal democratic principles and institutions of the nation—carefully articulating the role of science education within that general framework. Education must create (1) politically capable citizens with civic knowledge, skills for deliberation, and the ability to vote, serve on juries, and complete military service; (2) economically fit persons who are able to participate, compete, and make choices in economic life; and (3) liberal individuals capable of choosing their own ends. Given the lack of consensus on autonomy in the American political tradition, my account of liberal education emphasizes the demands of *pluralism* rather than the stronger requirements of autonomy. The liberal state is justified in securing a science curriculum—in terms of content and structure—that encourages students to develop skills that maintain democratic government and foster individual choice and freedom. This chapter uses *Plyler v. Doe*, education cases at the state and federal level, and Justice Stephen Breyer's *Active Liberty* to articulate a jurisprudence that deploys science education as a tool to develop citizens, reinforce liberal norms and general constitutional principles, and actualize the exercise of enumerated rights such as freedom of speech and press. Applying the logic of *Plyler* to Intelligent Design shifts the focus from identifying ID as religion to defining the relationship between science education, democratic institutions, and flourishing individuals.

CONSTITUTIONAL SILENCES

The Constitution is, with one exception, silent on science. Article I, section 8, clause 8 empowers the Congress to "promote the Progress of Science and useful Arts, by securing for limited Times to Authors and Inventors the exclusive Right to their respective Writings and Discoveries." In 1787, delegates to the Constitutional Convention suggested language that would have encouraged university teaching of science or other explicit supports of science, but none of these proposals were successful.[2] The Supreme Court has never developed a jurisprudence of science promotion based on clause 8 except the protection of patents and ideas.

The U.S. Constitution is equally silent on education. Charles Pickney's 1787 draft empowered Congress to "establish & provide for a national University at the Seat of the Government of the United States," but there is no record of a substantive debate on this clause, and no such language was included in the final draft.[3] Outside the Convention, Thomas Jefferson explicitly connected education and the stability of popular government when he maintained that education—particularly of working people with no access to private tutors or English boarding schools—limited the power of tyrants and aristocrats, helped maintain religious freedoms, and encouraged meritocracy.[4] As the drafting delegates never explicitly debated the relationship between education and republican government, the final document included no educational goals or institutions.

How might we explain these silences? Jefferson's vision of a republic powered by a discerning and engaged democratic citizenry lost out to James Madison's preference for *institutional* mechanisms that limited government power and preserved individual liberties such as the separation of powers and checks and balances. Jefferson's system depended on a population educated to be reliable participants in democratic citizenship, while Madison preferred indirect participation of the populace through the election of representatives.

The absence of established public school systems in the late eighteenth century also helps explain this constitutional silence on education. When the delegates composed the Constitution in 1787, most colonial children were educated at home, in church, or (for boys) at work through an apprenticeship. Families with the means sent their children to dame schools (a widow or childless wife teaching letters), English schools (reading, writing, and arithmetic), Latin grammar schools (preparing for entrance to college), or, in more populous areas, private-venture schools (teaching skills such as navigation and surveying). Since the seventeenth century, it had been common practice for towns in Massachusetts to contribute financially to schools in order to admit the poorer students, and this focus on paupers, orphans, and the poor was common throughout the colonies.

Although education at home was the norm at the time of the Constitution, some states had experimented with some form of public schooling. By 1671, Massachusetts Bay, Connecticut, Plymouth, New Hampshire, and Maine introduced legislation to encourage the establishment of schools. The most innovative state, Massachusetts, as early as 1647, had required towns with at least 50 families to establish an elementary school, but enforcement was lax. Pennsylvania's Frame of Government (1682) called for some publically funded schools to encourage reading (as well as reading of Scripture) and writing with explicit provision for education to allow the poor to have a useful trade or skill in order to support themselves through work. Virginia guaranteed some money to private Latin and English schools, but most of the funding targeted paupers and orphans—a model followed by many southern colonies. By the 1790s, the urban poor in most regions had access to free charity schools founded and funded by elites with the explicit aim of benefitting society by giving the poor a means of self-advancement. By 1800, half of the states had constitutional articles calling for public aid to education with an emphasis on education to ensure republican government (education as necessary to preserve rights and liberties) and maintain order as well as to enlighten minds and purify morals.[5] Although colonial and state constitutions outlined educational goals or provided for some schools, the laws guiding education at the close of the eighteenth century were inconsistent among the states.[6]

Nationally, a 1785 land ordinance law (drafted by Thomas Jefferson under the Articles of Confederation) called for a survey of the northwest territories and required townships to set aside land for local schools—land later converted to state land-grant universities. According to the Northwest Ordinance, "religion, morality, and knowledge being necessary to good government and the happiness of mankind, schools and the means of education shall forever be encouraged." Yet the notion of a national or even a state public school system would have been alien to most of the constitutional delegates—many of whom wished for constitutional protection of state sovereignty and a limited national government.

Constitutional silence and support for state sovereignty left education— goals, institutions, and standards—to the states and localities. By the 1830s, northern states experienced a "Common School Awakening" with education reformers like Horace Mann calling for free, tax-supported schools for all students—not just the poor. By the 1840s, more northern students attended public, as opposed to private, schools. Graded classrooms (including free high schools), a more uniform curriculum, and professional supervision served as common goals for school reform throughout the century as northern schools moved to face challenges posed by postwar industrialization and immigration.[7] In the South, the challenges of educating an agrarian, rural, and dispersed population combined with racist resistance to black education to thwart the development of any meaningful public education. Following a series of slave insurrections in the 1830s, localities passed fines for educating slaves, which succeeded in shutting down access. At emancipation, 95 percent of southern blacks were illiterate. The fear of educating blacks—before and after the Civil War—limited education opportunities for poor whites as well. There were advocates for free public schools for white children in North Carolina and Virginia, and some states provided funds for educating poor whites, but fear of government intrusion and the hostility of planters to educating poor whites or any blacks blocked the establishment of a public school system. After the Civil War, ex-slaves and proponents of white education pushed for public schools—and a limited number of segregated schools were created. *De facto* segregation became *de jure* with the end of Reconstruction and the beginning of Jim Crow, and black southerners had no or meager access to education. By the end of the nineteenth century, southern states had failed to create meaningful state standards, serve a majority of children, or provide anything approaching equity. In the South, school buildings, rules for attendance, and accessibility for immigrants and racial minorities varied significantly among the states.[8] Despite the expansion of schools in the South after the Civil War, southern white children spent fewer days per year and fewer years in school than northern children, even as the South spent "a larger share of its smaller economic base on its schools" than did the North.[9] During this period, northern blacks attended segregated schools, while popular fear of new immigrants fueled support for schools to Americanize them.

The Supreme Court remained silent on education throughout the eighteenth and nineteenth centuries—though their support of the doctrine of separate but equal in public transportation—*Plessy v. Ferguson* (1896)—was harnessed to support racial segregation in public schools. But by the turn of the century, public education was taking shape as a mass, national, institution that affected the lives of increasing numbers of Americans. While a small minority of teenagers attended high school in the 1890s, by the early twentieth century, high school had become a mass institution.[10] In 1870, 57 percent of 5–18 years olds attended school. By 1900, the number had jumped to 72 percent, in part, due to compulsory education laws that had been passed in all states by 1918 and enforced seriously by the 1930s.[11]

Beginning in the 1920s, the Supreme Court recognized American public education as an institution central to the lives of modern citizens through a series of rulings that coincided with the transformation of public education into a mass institution, as well as the related changes in the status of the United States as the world's leading industrial and military power.[12] Although states and localities remained the chief arbiters of goals and content, the Court reserved the right to determine whether an individual state's education policy was consistent with *explicit* constitutional provisions, such as equal protection or free speech. In the course of deciding whether states could forbid the teaching of German (*Meyer v. Nebraska*, 1923), segregate students by race (*Brown v. Board of Education*, 1954), require the reading of the Bible (*Abington School District v. Schempp*, 1963), mandate school attendance beyond the eighth grade (*Wisconsin v. Yoder*, 1972), or forbid certification of a non-citizen teacher who had not applied for citizenship (*Ambach v. Norwick*, 1979), the Court weighed the relative importance of access to public schools (to individuals and the nation) with the individual rights of students, parents, and teachers. The Supreme Court affirmed education as necessary for creating "self-supporting" and "law abiding" citizens who could "discharge the duties and responsibilities of citizenship," fulfill "social responsibilities," prevent tyranny, foster the "general welfare of society," and be "self-sufficient members of society." Education, the Court insisted, promoted "civic virtues," prepared individuals to participate as citizens, and preserved the democratic "values on which our society rests."[13] By the late twentieth century, the Court acknowledged a "compelling state interest" in universal education to "prepare citizens to participate effectively and intelligently in our open political system if we are to preserve freedom and independence" as well as to prepare "individuals to be self-reliant and self-sufficient."[14] And yet, even as the Court emphasized the duty of the state to protect "children from ignorance," it also reaffirmed that education mandates may not impinge on the fundamental rights of individuals.[15]

While the Court has never required the teaching of any particular *content* (e.g., science or scientific method), the Court has insisted that education is a private and public good that undergirds the political, economic, and personal life of the individual and underpins the liberal democratic principles and institutions of the nation. Instead of pursuing a strategy of proving that Intelligent Design violates the Establishment clause, I recommend instead that liberals and supporters of science education show that teaching evolution dovetails with this existing jurisprudence for citizenship, economic fitness, and independence by using a precedent from the late twentieth century.

FORGING A JURISPRUDENCE OF CITIZENSHIP: *PLYLER V. DOE*

In 1982, the Supreme Court ruled on a Texas statute that barred the children of non-citizens from attending the public schools. In *Plyler v. Doe*, the Court held—by a vote of 5 to 4—that the Equal Protection clause of the Fourteenth Amendment established the children as "persons" who cannot be discriminated against unless a substantial state interest can be demonstrated.[16] Justices Brennan, Marshall, Blackmun, Powell, and Stevens ruled that Texas could not deny public education to school-age non-citizens.

Justices Burger, White, Rehnquist, and O'Connor dissented. Immigration and naturalization appear unrelated to science education, yet the *Plyler* majority (Brennan) and concurring (Marshall) decisions articulate *why* education matters in a constitutional democracy and serve as an appropriate foundation for a liberal democratic jurisprudence of science and education.

Justice William Brennan's majority opinion justifies public education on two grounds: the collective impact of educated persons on basic American institutions (political and economic citizenship) and the individual impact on the life of the child as a potential adult (economic independence and liberal choosing). Both constructions map onto science education if scientific illiteracy undermines national security and diminishes economic competitiveness while limiting the economic, political, and personal choices of the individual.

William Brennan, appointed by Eisenhower in 1956, began his career on the Warren Court. He opposed the death penalty, supported abortion rights, and authored the "one person, one vote" majority opinion in *Baker v. Carr*. At his retirement in 1990, he was one of the remaining—and staunchest— liberal justices on the Supreme Court. For Brennan, American public education preserves the basic institutions and values of republican government. The public schools are the "most vital civic institution for the preservation of a democratic system of government," and the "acquisition of knowledge" is of "supreme importance" to the American people.[17] Public education supports democracy as the foundation of the successful performance of the "most basic public responsibilities," including military service.[18] But education is not just a means by which citizens are trained for discrete activities (like military service or jury duty). "Some degree" of education prepares children to preserve "freedom and independence" by participating "effectively and intelligently" in an "open" political system.[19] Thus, the nation relies upon education to train citizens to (1) perform specific civic duties and (2) preserve democratic institutions and the social order by fostering thoughtful and effective participation in the political system. Education has a third collective benefit: the socialization of potential citizens to the "shared values" of democratic government required for political stability and the maintenance of democracy. When education transmits and reproduces the values on which society and democratic institutions depend—and Brennan is confident, based on social science research, that it does—education contributes to the maintenance of the "social order and stability."[20] The denial of public education threatens the nation's ability to sustain its political and cultural heritage.[21] Stressing all three collective goods, Brennan concludes that leaving any group uneducated—particularly the children of non-citizens—threatens the "fabric" of democratic society.

Although Brennan highlights the collective benefits to society of public education, he does not ignore the private benefits to each child. Public education is not an individual Constitutional right, but "neither is it merely some governmental 'benefit' indistinguishable from other forms of social welfare legislation."[22] Illiteracy is "an enduring disability," and "education prepares individuals to be self-reliant and self-sufficient participants in society."[23] The uneducated individual suffers in all aspects of his life:

> The inability to read and write will handicap the individual deprived of a basic education each and every day of his life. The inestimable toll of that deprivation on the *social, economic, intellectual, and psychological well-being of the individual*, and the obstacle it poses to individual achievement, make it most difficult to reconcile the cost or the principle of a status-based denial of basic education with the framework of equality embodied in the Equal Protection Clause.[24]

As Brennan considers the "lasting impact" of education on the "life of the child," human achievement is an end in itself.[25] The Constitution does not require equality of *outcomes* (all individuals succeed at the same levels), but the framework of equality embedded in the Constitution's Equal Protection clause supports education as the prerequisite for equality of opportunity and self-reliance.[26]

Brennan might have followed Justice Douglas's famous dissent in *Yoder*. Asked whether the Amish can ignore Wisconsin's mandatory education requirements for high school students, Douglas dissented. The Amish had objected that high schools tend to "emphasize intellectual and scientific accomplishments"[27] that counter the religious traditions of the Amish. But Douglas countered that children must be educated long enough to evaluate their birth communities—and have the capacity to exit. Without sufficient education, the minors may not freely exercise religion (the foundation of pluralism) and are likely to blindly accept the choices of their parents.

Brennan prefers a narrative of individual achievement attached to a modified theory of human capital. Enhancing personal and professional choices and opportunities through education certainly benefits the individual, but Brennan suggests that self-reliance—especially economic fitness—has collective benefits:

> [e]ducation provides the basic tools by which individuals might lead economically productive lives *to the benefit of us all* . . . education has a fundamental role in maintaining the fabric of our society. We cannot ignore *the significant social costs borne by our Nation* when select groups are denied the means to absorb the values and skills upon which our social order rests.[28]

Education allows the child to compete in the economic marketplace—to "succeed in life"[29]—as it simultaneously creates a skilled labor force, social order, and democratic merit-based opportunity. Education also allows the "disfavored group" to "raise the level of esteem in which it is held by the majority."[30] After we cringe at Brennan's implication—we educate the children of non-citizens so that they will feel more approval from the majority—we recognize that Brennan's emphasis on the collective benefits of education may be strategic. As he defends public education for an unpopular minority group—the children of non-citizens—he signals concern with prejudice and status by appealing to the collective benefits to the majority rather than the individual benefits to the unpopular group.

Yet Brennan's education-as-human-capital argument still resonates, and we can apply it to science education. For example, a 2012 Council on Foreign Relations task force report argues that American educational failures threaten economic growth and competitiveness, physical safety, intellectual property, global awareness, and national unity and cohesion.[31] The under-education of American students leaves citizens without the "skills" or the "knowledge" necessary for modern industry, military service, and innovation; thus, the *nation* is at risk. Because, according to the report, science is one of the "building blocks for everything else," deficiency in science education—especially the inability to understand or apply scientific principles—is seen to contribute to all five problems.[32] Students are unable to "recall, interpret, critique, or evaluate texts. . . . [as well as] identify or use scientific principles in physical, life, or earth and space sciences, and they have failed to grasp science essentials such as the scientific method and inquiry-based learning."[33] The Council's report emphasizes national security rather than the flourishing liberal individual.

For theorists who privilege autonomy as an end in itself, the individual should build herself, not the national economy.[34] As discussed in Chapter 2, Harry Brighouse characterizes the human capital approach as "less crime, less rude," even as it contributes to civic education. Even as he acknowledges its collective benefits, Brighouse wants education to help the individual: "We exercise our powers of self-control, of rational thought, and of altruistic concern by being good citizens, and we also earn the respect of others; these things are genuinely valuable to us as well as to them."[35] Thus, economic self-sufficiency should be the means—not the end—of a rewarding life.[36]

Is Brennan's human capital approach too instrumental? Does it adequately protect the liberal individual's ability to make choices and flourish? Certainly, Brennan could further elevate the vital role of education in ensuring individual well-being and freedom, but *Plyler* balances the collective public benefit of an educated citizenry (maintenance of liberal democratic institutions, liberal citizenship, meritocracy, economic stability) with the individual private benefit (private and professional achievement; economic fitness; social, economic, intellectual, and psychological well-being; intellectual development; and human flourishing). Despite Constitutional silence on education, Brennan maintains that its importance in twentieth-century America—to the individual and the nation—cannot be questioned. Rather than suggest that education is an individual right, Brennan insists that public education is now an institution to which equal access results in both individual and collective good.

Brennan's constitutional confidence in equal access does not include prescribing the curriculum or methods of teaching that would create these politically, economically, and intellectually fit citizens. Following the Court's traditional deference to the states in the area of education, *Plyler* leaves the content and methods of public education to state laws and local governance.[37] Similarly, in *Brown v. Board of Education*, Chief Justice Earl Warren rejects segregation of children by race because it inevitably promotes inequality among citizens, but Warren does not specify *curriculum* that would encourage equality. Brennan forbids the exclusion of non-citizens and establishes broad goals, but he never identifies particular institutional or curricular mechanisms to encourage democratic socialization or individual self-development. He enumerates particular functions (like military service) for which an educated populace is necessary, but he never specifies values or principles to be inculcated in students.

Justice Thurgood Marshall's concurrence in *Plyler* also refuses to stipulate curricular content, but Marshall offers a more radical defense of education as a fundamental right—a position that has never been accepted by a majority of any Supreme Court. Marshall, the first African American to serve on the Supreme Court (1967–1991), argued *Brown v. Board of Education* before the Court as an attorney, and his opinions emphasize individual liberty in areas such as criminal procedure, capital punishment, and voting rights. His approach in *Plyler* (and an earlier education case, *San Antonio Independent School District v. Rodriguez*) helps us build an alternative jurisprudence based on his claim that education *directly* impacts enumerated constitutional rights and institutions. The "close relationship between education and some of our most basic Constitutional values" leads Marshall to conclude that education directly affects the individual's ability to exercise *particular* Constitutional functions of liberal citizenship: free speech, association, and the right to vote.[38] For Marshall, these individual rights support collective goals, such as open and informed debate on public policy. Unlike Brennan, however, Marshall does not emphasize economic achievement and national prosperity. Instead, Marshall channels John Stuart Mill's classical liberal argument for free speech as crucial to stable popular government, arguing that a system of "[c]ompetition in ideas and governmental policies is at the core of our electoral process and of the First Amendment freedoms."[39] Functioning as a marketplace of ideas, the classroom instills interest in "political discourse and debate" and encourages political consciousness and participation. Linking the marketplace of ideas to First Amendment protections, Marshall recognizes education as the backbone of constitutional institutions (elections), enumerated rights (freedom of speech, association), as well as democratic culture (discourse and debate on public policy, political consciousness, and participation). Education instills the *desire* to participate in political debate and appreciation for political principles as well as processes.[40]

Unlike Brennan, Marshall specifies the mechanisms by which education supports democracy. Marshall maintains, for example, that education increases the likelihood of voting.[41] The political franchise "has been afforded special protection" because it is "preservative of other basic civil and political rights."[42] Using social science research, Marshall emphasizes the direct relationship between education and voting in the 1968 presidential election—and his view is supported by modern studies, which consistently demonstrate that educated citizens are more likely to vote in all elections and primaries.[43] Although the crucial leap in voter participation takes place when students finish college, higher education levels—including the completion of high school—continue to predict the likelihood of voting. If the constitutional system relies on elections to select two branches of the government, education that encourages voting becomes a necessary (though not sufficient) condition for good government. Brian Barry counters that making a "good job of voting in elections" is in the interest of the whole, not the individual.[44] Certainly, the vote of an individual rarely determines an election, but Marshall seems to have something else in mind. An electorate that can detect fallacies or question policy is an asset to the nation, but the *act* of voting leaves the individual with a sense of equal citizenship, respect, and ownership of the democratic process that Marshall values independently.

Here, Marshall appeals to both "specific Constitutional guarantees" (liberal citizenship and collective benefit) and "particular personal interest" (individual development and human flourishing).[45] Education enhances the ability of individuals to speak openly, voice new ideas, and criticize the government. Thus, free speech and association empower individuals *qua citizens* to check government power while simultaneously enabling them to grow as individuals. A student inquires, studies, and evaluates "to gain new maturity and understanding" that he will "enjoy throughout his life."[46] The individual's enjoyment of her rights may be enhanced by education.[47] In his *Yoder* dissent, Douglas held that a child without education is "forever barred from entry into the new and amazing world of diversity that we have today." If the child is only exposed to the beliefs of the Amish, "his education is truncated," and "his entire life may be stunted and deformed."[48] Marshall does not fully endorse Douglas's position on individual autonomy, but he endorses rights as *both* contributing to the public good and individual "maturity." The ability to determine what is valuable through critical reflection and experimentation empowers the individual, in addition to enhancing collective political or economic utility.[49] Education encourages citizens to challenge values and inherited traditions for the sake of a healthy critical democratic culture as well as the individual's self-development. "Governing ourselves" has two meanings. Collectively, the people rule through their representatives. Privately, humans manage personal decisions without overbearing intrusion from government, groups, or other individuals. Both forms of self-governing require information, the capacity to exercise judgment, and free speech rights.[50]

MAPPING SCIENCE ONTO CITIZENSHIP

Neither Brennan nor Marshall provide a rule for determining what must be included or excluded from public school curricula, but the logic of *Plyler* changes the questions that the Court might ask with regards to Intelligent Design. Instead of asking whether ID is religion that establishes religion, the court might *also* ask whether ID hinders democratic institutions or individual flourishing? With different emphases, Brennan and Marshall argue that education advances citizenship, economic fitness, and individual self-development. They offer modified human capital models in which individual choice and achievement affect wider political, economic, and social institutions without specifying the content of the education that would produce such citizens, workers, and persons. A growing set of state cases connect education, civic capacity, and preparation for employment—like *Plyler*—but these cases—unlike *Plyler*—highlight the role of scientific literacy. For example, the New York Court of Appeals held that education should create the following: (1) *voters* who have the "intellectual tools to evaluate complex issues, such as campaign finance reform, tax policy, and global warming," and the capacity to vote; (2) *jurors* who could "determine questions of fact concerning DNA evidence, statistical analyses, and convoluted financial fraud"; and (3) *citizens* who could economically support themselves, so that they were not dependent on the state for resources, and could help expand the economy.[51] Voters and jurors who must understand global warming or DNA evidence must have some basic scientific literacy, and the National Academy of Sciences (NAS) frequently emphasizes the link between the theory of evolution and the basic assumptions of DNA.[52]

A similar—yet more direct—argument for science education as essential to democratic citizenship can be found in two *amicus curiae* in *Edwards v. Aguillard*. In *Edwards* (discussed in Chapter 2), the Supreme Court found Louisiana's Balanced Treatment for Creation-Science and Evolution-Science Act (which required teaching creation-science whenever evolution was taught to students) to be an unconstitutional violation of the Establishment clause. Justice Brennan wrote the majority opinion in *Edwards*—just five years after *Plyler*—but he made no appeal to education as constitutive of democratic citizenship. He also made no mention of the arguments in two briefs—submitted by scientists and scientific organizations—that made an argument similar to his own argument in *Plyler*. Examining these two briefs helps flesh out the argument for a citizenship jurisprudence (based on *Plyler*) that explicitly includes science education.

The first *amicus* brief, submitted by 72 Nobel laureates in physics, medicine, chemistry, or physiology as well as 17 state academies of science and several national organizations, such as the American Anthropological Association and the Association of American Medical Colleges, argued that teaching creation-science alongside evolution damaged national science education, thus, the capacity of the nation to "cope with problems of food production, health care, and even national defense."[53] Good public policy, national security, and "our ability to respond to the problems of an increasingly technological world" would be "jeopardized if we deliberately strip our citizens of the power to distinguish between the phenomena of nature and supernatural articles of faith."[54] Like the New York Court of Appeals and *Plyler*, the 72 Nobel laureates brief connected scientific literacy to education for citizenship, work, and individual deliberative capacity.

A second *amicus* brief submitted by the NAS emphasizes individual intellectual development. If science education preserves natural curiosity and inquisitiveness as well as the capacity to imagine new solutions to problems, the NAS contends that science education shapes "immature minds." Scientific inquiry—with its emphasis on "rigorous and systematic methods of observation and evaluation of empirical data"—helps supply "the intellectual tools needed to enter adult society." Once students understand what science is, how the scientific method operates, and what science seeks to accomplish, they more fully understand the natural world. Undermining science education threatens to "stunt the intellectual development of generations of American children."[55]

For NAS, science not only enhances general intellectual development, it specifically empowers students who pursue careers as scientists, engineers, or other occupations requiring "technical training and skill." But NAS is even more concerned with the relationship between individuals who choose science and the public good. The *nation* requires "a scientifically literate citizenry" as well as a large pool of scientists, technicians, and engineers. The national need for such professionals justifies compulsory education and tax-supported public schools, but the justification loses force when the state (Louisiana) intervenes to inject creation-science—or any other discredited science—into the curriculum. Individual intellectual development as well as national needs for professionals are thwarted.[56]

The NAS links both goals—a scientifically literate citizenry and a flourishing pool of scientists—to the academic freedom of the high school science teacher. The Louisiana statute compels a teacher who concludes that evolution reflects the best available science to nonetheless teach creation-science. Teaching material that has been found to be without value as science by the scientific and academic communities may confuse, misinform, or manipulate students—and violate the academic freedom of the teacher. The First Amendment, says the NAS, should allow the teacher to teach evolution—and be silent on creation-science. The law at issue in *Edwards* would be thwarted by such an approach because it *required* teachers to teach both. But can this First Amendment claim defend against the new forms of legislation that give teachers the freedom to introduce ideas that they believe are true (e.g., creationism or Intelligent Design)? A teacher who favors ID—according to the NAS logic—should be allowed to teach her scientific beliefs in her own classroom. The new Tennessee legislation capitalizes on this protection of academic freedom.

The NAS has a response: evolution is not a *belief*, and creation-science is not *science*. Like the U.S. District court in *McLean* (see Chapter 1), the NAS rejects teaching information that is "not science" in a science curriculum as well as material that is religious and violates the Establishment clause. The NAS—as well as the 72 Nobel Laureates—carefully outline a set of criteria for science. Academic freedom in a *science* class does not include expressing your beliefs—only scientific theories that have evidence to support them may be taught. Both briefs carefully and extensively define terms (science, law, theory, fact, empirical observation, naturalistic, hypothesis, *ad hoc* hypothesis) while contending that the Louisiana legislature and some courts are misusing the terms. The criteria for science is similar to those suggested in both *McLean* and *Dover*. Science is (1) naturalistic (rather than supernatural), (2) testable, (3) tentative (may be disproved by *empirical* observation), and (4) predictive and explanatory (can predict or explain phenomena in the natural world).[57] For the scientists and science organizations, basic science education required high school curriculums to "accurately portray the current state of substantive scientific knowledge" as well as the "premises and processes of science." Teaching anything that is "not science" invalidates both requirements.[58]

When faced with Louisiana's requirement to teach creation-science, the Supreme Court never returned to its own jurisprudence on education or the "not science" claim from *McLean* or the *Edwards* briefs. Why not? First, it is difficult to formulate a constitutional question. Does the Louisiana act violate academic freedom as free speech under the First Amendment made applicable to the states through the due process clause of the Fourteenth Amendment? Maybe. But Intelligent Design supporters can claim *their* academic freedom is violated when they cannot express themselves in the classroom, and, in his *Edwards* dissent, Justice Scalia accepted the "data" of a tiny number of rogue scientists in order to conclude that creation-science was a credible *scientific* alternative, not a religious belief. Because the NAS first amendment argument relies on a reliable source for best available science, a first amendment argument requires the Court to adopt criteria for science (similar to *McLean* or *Dover*).

An alternative constitutional question would be, "Does presenting unsubstantiated opinion violate/impede Article I, section 8, clause 8 that empowers Congress to 'promote the Progress of Science and useful Arts, by securing for limited Times to Authors and Inventors the exclusive Right to their respective Writings and Discoveries' "? Certainly the NAS believes that generating inventions and encouraging scientific/engineering discovery partially justifies public school science education. The Court could interpret the protection of patents to reflect the intent to protect (and encourage) scientific discovery, but there is no established jurisprudence here linking patents and education.

What would a science and citizenship jurisprudence based on *Plyler* look like, and how might it address constitutional silences on both education and science? Neither Brennan nor Marshall provide a rule for determining what might be excluded from the curriculum, but both assume that public education undergirds citizenship, economic fitness, and individual decision-making. Here, they are following an established jurisprudence of citizenship, economic fitness, and individual growth, but the Court has always invoked this education jurisprudence in response to questions involving other parts of the Constitution (e.g., free exercise in *Yoder*, equal protection in *Plyler* or *Brown*).

The Court cannot credibly claim that the theory of evolution is the foundation of citizenship, but it can suggest that the *constitutional structure* and specific rights depend upon individuals who vote, serve in the military, sit on juries, and assemble to resist tyranny (an educated citizenry), and scientific literacy contributes to the flourishing of democratic institutions, economic fitness, and individual self-development. Thus, teaching science that has been discredited threatens the creation of an informed citizenry and liberal choosers. The argument would emphasize both the collective and individual benefits of education, and the Court would steer clear of any argument based on values. The emphasis would be on skills and abilities necessary for a particular regime.

For example, the Court could offer an educational penumbra similar to Justice Douglas's famous defense of privacy in *Griswold v. Connecticut*. In order to reject Connecticut's ban on the use of contraceptives for married couples, Douglas claimed that the Constitution protected privacy through a penumbra—rather than an explicit clause—and he identified the First, Fourth, Fifth, and Ninth Amendments. A democratic education penumbra would include free speech and academic freedom (First Amendment), the capacity to vote (Fifteenth, Nineteenth, and Twenty-Fourth Amendments), the right to raise and support a military force (Article I, section 8, clauses 12, 13, and 14), and the right to a jury trial (Article 3, section 2, Sixth and Seventh Amendments).

This jurisprudence of citizenship and education is similar to Justice Stephen Breyer's liberty and citizenship jurisprudence in *Active Liberty*.[59] For Breyer, the Constitution embodies two principles of liberty. "Modern" liberty protects individuals from government by supporting the rule of law, ensuring that government is democratic, avoiding the concentration of political power, protecting personal liberty, and insisting that the law equally respect each individual.[60] Political theorists call this negative liberty. "Active" liberty guarantees that citizens are free to participate in government. Citizens are sovereign when they govern themselves, and they are the source of political power.[61] For Breyer, popular sovereignty requires informed participants who can contribute to a national, democratic conversation and help create sound public policy.[62] Because citizens actualize Constitutional institutions and principles, Breyer obliges courts to guarantee citizens both modern and active liberty. The logic of the Constitution requires a particular type of citizen to realize institutions and values.

Breyer isn't quite right about active liberty because he relies on questionable assumptions regarding ancient liberty and doesn't provide clarity about positive liberty claims. Yet his most important claim is correct: the logic of the Constitution requires a particular type of citizen and person to realize institutions and principles. The most constitutionally consistent approach to ID is to think about citizenship and education not in terms of values (because liberals are understandably nervous about inculcating values) but in terms of skills and information required for political citizenship. In the twenty-first century, scientific literacy is part of that skill and information set.

When we connect Breyer's conception of the Constitution and *Plyler's* understanding of education, we radically reframe the debate over teaching evolution in public schools. Breyer supplies a rationale for requiring citizens capable enough to operationalize the Constitution, but he does not elaborate *how* modern or active liberty can be created outside of Supreme Court rulings. *Plyler* provides a broad argument for education as the foundation for citizenship and a motivation for participation. If we employ an active citizenship (rather than liberty) jurisprudence, courts can ask whether teaching the best available science furthers political citizenship, economic fitness, or individual self-development.

CONCLUSION: A CONSTITUTIONAL EDUCATION?

The majority decisions in both *Edwards* and *Dover* do not treat teaching creation-science or Intelligent Design in terms of the wider Court jurisprudence on education. Applying the logic of *Plyler* to *Dover* shifts the focus from defending against ID as religion to defining the relationship between science education and citizenship. I am not suggesting that the Supreme Court can—or should—single-handedly change curricular goals. Realistically, the current Court might not uphold the rulings in *Plyler* or *Edwards,* and curriculum remains the domain of the states. Yet the Court—and others—should consistently and aggressively insist that the logic of citizenship in the Constitution requires education to shape civic-minded, participatory citizens who perform their duties, participate in public policy debates, achieve economic independence, and make private decisions toward moral personhood. The *Plyler* logic provides an opportunity to go beyond the defensive establishment clause jurisprudence and explicitly engage the public and private aspects of science education.

The Constitution's silence regarding education, values, and citizenship leaves the courts laboring to define ID as religion that can be excluded. The citizenship and education jurisprudence forwarded in this chapter allows us to go beyond the inadequate language provided by *Dover* and *Edwards* to articulate *why* the teaching of evolution is essential to liberal democratic practice and institutions. Fusing the ideas of Brennan, Marshall, Breyer, and the science *amicus* briefs would create a viable alternative: a vocabulary capable of addressing questions of education and civic membership.

Following this logic, discredited science should be excluded from public schools because it harms the individual's capacity to flourish in public and private life and cripples society's capacity to respond to policy problems involving science and technology. The Court should decide the next case using the Establishment clause (because it is the settled precedent supported by clear text), but it should include an auxiliary argument based on the education cases—as well as the not-science criterion of *McLean* and *Dover*—in order to establish science education as relevant to constitutional requirements for citizenship, economic fitness, and liberal personhood. The emphasis should be on the skills and content necessary for self-governance, collectively and individually.

4 Science Education for Citizenship Skills

Dewey, Darwin, and the Democratic Temperament

The thought of John Dewey helps connect the Supreme Court's conception of the political citizen, Amy Gutmann's desire for deliberating majorities, Harry Brighouse's focus on skills, and teaching evolution in public schools. Stephen Macedo's fear of John Dewey as the bogeyman of positivism should not prevent liberals from using Dewey to explain why *science* education is essential to liberal democratic citizenship. Joining Dewey's focus on science with the skills approach suggested by Brighouse and Barry yields a principled and politically practical defense of liberal education.

Like Thomas Jefferson, Dewey believes an educated majority can be trusted to govern—and govern well. Science education is essential because it exposes students to a mode of thinking: the scientific method. Science is "that knowledge which is the outcome of methods of observation, reflection, and testing which are deliberately adopted to secure a settled, assured subject matter."[1] The scientific method is a form of intelligence: a method of experimental inquiry that demands the collection of facts and the observation of relationships between objects. Scientific attitude requires a mind disciplined to observe the world and understand causal connections. The essence of a scientific attitude is active questioning, delight in the problematic, and desire for a search.[2]

Dewey maintains that sixteenth-century European scientific discoveries led to a rejection of *telos*—the insistence on fixed tendencies toward definite ends, intentional activity, and defining forms—in favor of the observation of patterns provided by experimental inquiry and the operation of reason. Although the scientific method revolutionized Western Europe's understanding of the *physical* world, it left moral and political ideals virtually intact. While Dewey is clearly wrong—the scientific method affected the study and content of moral and political ideas in early modern Europe—Dewey correctly identifies Darwin as a revolutionary figure.[3] Before Darwin, the rejection of purpose in physics and chemistry co-existed easily with a continuation of *telos* as an explanation for the creation and changes in plant and animal life. Challenging the design theories of his age, Darwin exploded this balance by presenting a theory of animal and human origins that did not depend upon a supreme being.[4] The modern Intelligent Design movement, like earlier design movements, straddles *telos* and scientific method as it accepts elements of natural science *and* insists on the supernatural guidance of human evolution.[5]

For Dewey, the scientific method changed the West's understanding of the natural world, and, after Darwin, the method affected the *content* of political, social, and moral norms. This is the Dewey that gives Macedo and others pause: the extent to which Dewey claims that the method *determines* the substantive outcome. But Dewey's version of the scientific method focuses on revision—an element of critical inquiry that Brighouse and Brennan believe is part of liberal citizenship. For Dewey, science education does not merely advance our chosen goals. It forces us to question long-held beliefs—based on tradition, chance, dogma, custom, superstition, self-interest, personal prejudice, and class interests—and modify our ends. If properly taught, science leads democratic citizens to question traditional religious and political institutions and change the substance of their economic, political, and moral values and goals.[6]

Moreover, the scientific method enables the democratic public to recognize the nature of the problems it must control (e.g., comprehending the costs and seeing causal connections) and form a shared will. It also contributes to industry and economic development. Even as Dewey speaks to the collective benefit of scientific method, he understands that science education will develop each *individual's* capacity to think. If every citizen has a voice in decision-making, these capacities can support democratic deliberation and public reason. As humans rely on scientific explanations rather than mystical, religious, or superstitious explanations for natural phenomena, they come to control themselves and gain "intellectual self-possession."[7] Thus, the healthy functioning of democracy, economic development, and individual self-governance are inextricably linked.

Appealing to Dewey has well-known baggage: the inclination toward mastery of nature, the naïve investment in progress, and the support of science as an absolutist philosophy or a "faith."[8] Nevertheless, avoiding Dewey has costs because his conception of education and the scientific method clarifies what is at stake in today's debate over Intelligent Design and the teaching of evolutionary theory. Liberals should reject positivism or any suggestion that science is an objective arbiter of social problems while they embrace Dewey's explicit defense of *science* education as useful for democratic deliberation. Dewey's democratic temperament includes the ability to distinguish good scientific arguments from bad ones and requires citizens who can weigh facts, observe relations, and reason carefully. His definitions of science and his emphasis on questioning and searching recall arguments made by the National Academy of Sciences, the courts in *McLean* and *Dover*, as well as Brighouse and Barry.

Returning to the logic of *Plyler* after considering Dewey's claims for a critical temperament through education helps us create a liberal education discourse focused on science education. Science for citizenship claims that liberal democratic governments need politically capable citizens with civic knowledge, deliberative competence, and the ability to vote, serve on juries, and complete military service. They also require economically fit citizens who are independent, self-reliant, and able to participate and compete in the economy. As such, liberal education must encourage competencies that enable Constitutional rights, duties, and institutions. Because citizens disagree about values and the good, liberal education should not emphasize values, autonomy, or virtues. Instead, liberals and modern courts should frame liberal education in terms of the skills necessary for self-governance in two senses: individuals collectively governing the nation through democratic action and the exercise of rights and the individual governing herself in private life. Using the language of *Plyler* and resources from contemporary liberal theory, liberals should begin to explain how science education affects the development of a democratic temperament, including the critical thinking capacities that nourish liberal citizenship, economic fitness, and ethical decision-making.

What are the capacities essential for citizenship? No list will be perfect, but these skills include resisting tyranny, exercising rights (particularly those linked to dissent and criticism), and making independent choices in economic, personal, and political life. Democratic education should encourage skills and capacities relevant to these ends. At bottom this will include the capacity to evaluate evidence, policies, candidates, culture, and traditions in order to function as citizens, jurors, consumers, soldiers, workers, or moral persons. The capacity to evaluate must include the ability to detect fallacies in arguments and acquire and weigh evidence. Citizens must be able to criticize and challenge authority, including avoiding exploitation and dealing effectively with public officials.[9] In public, citizens should be able to understand and engage the arguments of fellow citizens, and this may require imagining themselves in the position of others in order to understand and tolerate other points of view and consider those points of view when making choices and policy.[10] Although the democratic polity cannot guarantee that all citizens will be equally proficient, democracy requires a citizenry with some combination of these aptitudes to deliberate in public life.[11]

Economic fitness is embedded in this vision of the democratic citizen. Students must be educated to make choices in the economy—to seek, choose, hold, and change jobs—and to do so legally.[12] Individuals, regardless of their private goals and desires, need employment to fulfill basic humans needs (food, shelter, clothing), enjoy leisure, and (in some cases) experience work that is personally interesting or fulfilling.[13] Brian Barry includes making a living legally, engaging in commercial transactions, and avoiding exploitation (economic or political) due to ignorance or incompetence.[14]

In the debate over evolution education and Intelligent Design, focusing on skills and capacities—rather than values and autonomy—allows liberalism to be respectful to the challenge of pluralism as well as the dangers of state power. As liberalism struggles with diversity and assimilation, we are faced with a difficult balancing act: "the importance of reproducing and improving our liberal democratic political community" versus the "private interests of individuals and families."[15] Public schools can require that students learn critical thinking, but they cannot force them to act autonomously or to favor individualism rather than community. Because education is a powerful institution, liberal education should be cautious, and the state should not coerce children into a particular belief system. But this caution need not be spineless. Pluralism should be understood as a choice made by an individual—the student who will become an adult—rather than the parents, families, or communities that the children did not choose. Real pluralism requires real choice, and liberal education should train students to think and evaluate.

A focus on skills and capacities for a democratic temperament cannot, however, ignore content, and, given the current political climate, it cannot ignore issues of science education. As liberals argue that *all* disciplines contribute to students' ability to weigh evidence and think critically, they must overcome their discomfort with positivism to explain how science education—both scientific method and content knowledge—help build these capacities. Liberals must address two issues: teaching content is not positivism, and science is substantively part of the liberal education that encourages critical inquiry.

Positivism has threatened the integrity of political science as well as political theory—and theorists should be wary, but we cannot allow this to make us timid about answering literalist objections to best available science. As we reject science as a neutral standpoint—as capable of answering questions—we must replace it with an understanding of science as a tool for making choices in political, economic, social, and ethical life. The emphasis must be on science education—method and content—as providing knowledge relevant to public discussion. The first step is to establish the relationship between evaluation, deliberation, and content knowledge. Martha Nussbaum's work on the humanities provides insights. Although Nussbaum never fully explores how science fits into a liberal arts education, she sees the need to have the data to evaluate specious assumptions—like the "melanin theory of culture"—as well as to make personal decisions that rely on such scientific knowledge. Nussbaum supplies the example of a gay student at Brigham Young University ignorant about basic human sexuality. His lack of information—and the presence of *dis*information (e.g., homosexuality does not exist in nature)—affect his choices, specifically, the ability to evaluate and exit a community.[16] Although Nussbaum does not connect her capabilities approach to science, science education would affect the individual's ability to critically examine herself and her traditions as well as formalize arguments, detect fallacies, reason logically, justify arguments, and test reasons for "correctness of fact, and accuracy of judgment."[17] Liberal persons need a complex sense of the world—as it actually is—to govern themselves and others.[18] Content knowledge allows citizens to assess ways of life, policy, politics, and candidates and contribute to what Brian Barry calls a "mutually intelligible discourse."[19] Science does not provide answers on policy, but science education helps shape citizens more capable of following a newspaper article on climate change or a juror better able to understand the reasons for excluding DNA evidence.

* * *

Any list will disappoint someone, and the emphasis on basic temperament and skills rather than civic education, autonomy, or liberal values will frustrate theorists who crave a thicker commitment to autonomy or liberal values. Skills are not neutral, for they enable a particular kind of citizen—and even learning skills will have non-neutral effects on communities. Yet skills are less coercive than values, and a narrow definition of citizenship—like the one provided in *Plyler*—may yield more agreement. Citizens should, at bottom, think critically about tyranny; observe problematic leaders, groups, and institutions; and act upon those observations through freedom of speech, association, and the franchise.

In the context of the struggle over teaching evolution, teaching science to develop skills essential to a democratic temperament allows liberals to argue *for* content or skills that develop citizens rather than objecting to Intelligent Design as religion. If a district or teacher proposes to teach Intelligent Design, the conversation should be about whether this is good science that can be proven through the scientific method (rather than objecting that it is religion that must be excluded). The focus should be on how accurate information develops a student's capacity to weigh evidence and govern the country and themselves. Science should never be offered as the only source of the skills necessary for democratic temperament—but it must always be part of the education that allows students to become citizens capable of public reason.

5 Earthquakes and Plows
The Eighteenth-Century Democratic Temperament

The self-governance jurisprudence—education is central to the self-government of the nation and individual—is based on the Supreme Court's education jurisprudence (90 years) and two critical Supreme Court decisions (*Plyler* and *San Antonio*)—as well as the Constitution's basic structure. Education for self-governance and a democratic temperament is also deeply embedded in American political thought.

This chapter situates the self-governance jurisprudence in the works of John Adams, Thomas Jefferson, and Thomas Paine, arguing that the American political tradition has a consistent narrative that links education, science, and citizenship—as well as a recurring tension: science as a threat to religion versus science as knowledge that can bring citizens closer to God. The works of Adams, Jefferson, and Paine illustrate the role of the educated citizen in the liberal political tradition: education is central to the development of a questioning temperament that protects rights and thwarts tyranny and dangerous elites. As science feeds the critical spirit necessary for the creation and maintenance of popular government, it may inspire citizens to scrutinize religious institutions, yet this narrative of self-government that celebrates science and discovery can be deeply religious.

JOHN ADAMS'S EDUCATION FOR LIBERTY: KNOWLEDGE, TYRANNY, AND THE POOR

In November of 1755, John Adams felt his family home in Braintree, Massachusetts, "rock and reel and crack as if it would fall in ruins about us." Adams concluded that the force was one of nature: a "very severe shock of an earthquake" that lasted about four minutes and shattered chimneys within a mile of his "father's house."[1] For other prominent colonists, the force was an act of God: a divine punishment for immorally lived lives and the failure to keep the faith of the Pilgrims and Puritans. In Boston, the emphasis on the divine took an extraordinary form. The Reverend Thomas Prince of South Church claimed that God sent the earthquake to punish Benjamin Franklin for inventing the lightening rod. Franklin's rod stopped

God from striking buildings and people with lightening. To show his anger, God sent an earthquake.[2]

A week after the earthquake struck, Harvard's John Winthrop responded to Reverend Prince. Winthrop, who had trained John Adams as a student, devoted a class lecture and subsequent pamphlets to the science of earthquakes.[3] The earthquake was a natural—not a divine—force. Adams dismissed Reverend Prince's charges, and Franklin asked why it was tolerable to build a roof to keep out the rain but unacceptable to divert the lightening? Winthrop, Adams, and Franklin assumed that God *had* created the universe and its forces, yet they insisted that science—rather than divine punishment for immorality—explained the earthquake. Understanding the natural order did not lead them to question God. For Reverend Prince, a scientific explanation of the earthquake diminished God's power and destabilized religious faith.

This eighteenth-century encounter helps us understand the current controversies over teaching science in public schools. From colonial times, Americans have split on how to reconcile religion and science. For Adams and Winthrop, belief in both God and the natural causes of earthquakes are compatible and mutually affirming. For Prince, science threatens to destabilize religious faith. Like Reverend Prince, opponents of teaching evolution insist that exposure to the content of the theory of evolution and the assumptions of natural science will lead students to question *God*. As we saw in Chapter 1, the exclusionary argument (belief in evolution means rejection of God) is a powerful political and rhetorical tool as most Americans report that they believe in God and affirm some form of theistic evolution. Unintentionally, the defensive approach to Intelligent Design (exclude ID as religion using the first amendment) leads to further confusion as the public assumes excluding creation or Intelligent Design means excluding *God* as opposed to rejecting discredited quasi-scientific claims.

Why was science education so important that Winthrop would devote pamphlets and public lectures to the science of earthquakes? Winthrop and Adams believed that free and stable government relied on a particular kind of citizenry—knowledgeable, critical, and questioning—and they did not separate scientific knowledge of the world from the study of philosophy or even theology. For them, knowledge of the world respected rather than threatened God. Adam ties this knowledge to tyranny prevention, the energy of rights, and the creation of "a people" based on liberal principles rather than cultural origins.

* * *

In the midst of controversy over the Stamp Act, John Adams read three papers before a group of Boston lawyers designed to place the Act in the perspective of the history of liberty.[4] Reprinted in the *Boston Gazette* and later known as the *Dissertation on the Feudal and Canon Law*, the papers argued that learning and knowledge are the foundation of liberal citizenship. In order for free government to be stable over time, citizens must be (1) capable of voting out tyrants and (2) able to exercise their rights to keep government smart. Adams believes education leads to this type of citizenry. Adams links education to the exercise of specific rights: speech and press.[5] While ignorance leaves the people timid and stupid—leading to "cruel tyranny"—education creates knowledgeable citizens who courageously act to prevent oppressive government.[6] Education prevents tyranny by helping the people both understand and exercise their rights. Knowledge allows the people—especially the poor—to collectively oppose tyrannical government. Education enables the "common people" to understand their rights—even if those rights are God-given or innate—and successfully reject mistreatment by those in power. Unlike Kant, Adams does not speak to the effects of education on the individual; instead, he emphasizes the benefits of education to the entire society because free government requires a change in the understanding of "the people."[7]

Adams acknowledges that the sheer cruelty of arbitrary government encourages the people to seek independence and confine the "power of the great within the limits of equity and reason."[8] Here, Adams relies on a classic liberal formulation: spontaneous self-interest propels the individual to resist the tyrant or the force of other individuals or groups. Adams assumes the logic of the social contract, but he believes it is insufficient to assume that people "know" the logic of the contract without *formal* education. The people need more than the *experience* of cruel government or elite dominance to successfully rebel. Although Adams consistently maintains that rights are God-given and cannot be repealed by human laws, the people benefit from education that explains the dynamic of rights: education enables and energizes rights.

In order to effectively oppose tyranny, the people need education in "arts and letters" and the leisure to pursue it. Attending to the role of class, Adams comfortably addresses the differences between the "great" and the "meanest ranks" of the people.[9] Assuming constant class tension, Adams explicitly seeks to empower the poor to be more vigilant citizens through education:[10]

> The poor people, it is true, have been much less successful than the great. They have seldom found either leisure or opportunity to form a union and exert their strength; ignorant as they were of arts and letters, *they have seldom been able to frame and support a regular opposition.*[11]

Because the "great" understand this dynamic, they keep the "knowledge of their rights and wrongs, and the power to assert the former or redress the latter" from the people. Education not only lessens the power of oppressive and arbitrary government, it empowers the poor to limit the power of elites. The ability of the nation to fend off tyranny depends upon the education of the poor in their rights and the responsibilities of government. Such knowledge lessens the power of oppressive and arbitrary government because the poor are better equipped to "frame and support" an opposition.[12]

Adams focuses on the need to diffuse knowledge "generally through the whole body of the people" and especially the "common people" or the "lowest ranks." Adams slides between "the poor" and "the people," but he insists that the education of the "lowest ranks" is of more importance *to the public* than the property owned by the rich.[13] Aristotle insists that the poor, collectively, have more property than the rich, but Adams's claim is more radical. The education of the poor is more valuable to the common good than the property of the rich. The function of education is *not* individual self-realization or equity for the poor. Instead, widespread education benefits the public and the "country's welfare." Adams claims that this heritage of education differentiates Americans from Europeans. An American who cannot read or write is "as rare as a comet or an earthquake," and "we are all of us lawyers, divines, politicians, and philosophers." Foreigners observe that "they have never seen so much knowledge and civility among the common people in any part of the world." Education helps forge a collective national identity of rights-bearing, tyranny-preventing, elite-limiting liberal citizens. The education of "the people" can be, therefore, justly secured by public expense.[14]

Although Adams focuses primarily on the education of the "people" and the poor, he believes elite "men of learning" helped prevent tyranny during the reigns of James I and Charles I. Thus, universal elementary education should be complemented by a college-educated elite.[15]

But educating widely and providing for elites is not sufficient. Adams recommends targeted education to enable citizens to understand and exercise *particular* common law rights: freedom of speech and the press. His argument mirrors Justice Marshall's dissent in his *San Antonio School District v. Rodriquez*. Although Marshall identifies speech and *voting* as the key rights, both men agree that education fosters *particular* rights that maintain free government. For Adams, knowledgeable citizens restrain the powerful because publishing the "speculations of the curious" preserves the "freedom of thinking, speaking, and writing" and reveals the vices of "great men."[16] A free press protects the public by publishing *and* fostering criticism. Rather than emphasizing the effects of publishing or reading on the individual's autonomy of self-expression, Adams focuses on the *collective* effect: free press serves the "country" because individual citizens thwart oppressive government.

In celebrating the freedom of the press, Adams returns to the theme of class. His hatred of the Stamp Act stems from his belief that an educated public—that will read and publish in newspapers—limits the power of the government and elites. The Stamp Act is particularly objectionable because it strips citizens of a "means of knowledge" and takes "from the poorer sort of people all their little subsistence."[17] Adams's focus on the poor's ability to recognize and thwart tyranny or buy newspapers makes schooling open to all even more important. Here, he creates the logic for free public education: the entire public must exercise rights. Adams's conception of citizenship asks a lot of citizens—but he insists that governments should provide the institutions necessary to create citizens capable of such a weighty responsibility: public schools.

Adams admits that there are Americans who resent providing universal education as "an imposition upon the rich in favor of the poor, and as an institution productive of idleness and vain speculation among the people, whose time and attention, it is said, ought to be devoted to labor, and not to public affairs, or to examination into the conduct of their superiors."[18] Nevertheless, he insists that the education of the poor supports *common* liberal goals, like preventing tyranny.

At this point, Adams appears to supply a secular justification for the education of the citizen: prevention of tyranny and elite control. But Adams conceives of citizenship in a disconcerting mix of religious stereotyping and egalitarianism. Puritans best embody the education necessary for liberal citizenship. Like many of his contemporaries, Adams writes with simplistic certainty about the tyrannical perils of Catholicism and the liberal assets of Puritan Protestantism. For Adams, the American Puritans resemble the first Protestant reformers in England: intelligent, learned, and "better read than even the members of the church." The American Puritans prevented the two systems of tyranny—feudal and canon (Catholic)—by educating the whole body of the people. According to Adams, they radically supported universal public education by requiring that all towns have grammar schools "[s]o that the education of all ranks of people was made the care and expense of the public, in a manner that I believe has been unknown to any other people ancient or modern."[19] Indirectly, Adams refers to the Massachusetts School Laws passed by the legislature of the Massachusetts Bay Colony in 1642, 1647, and 1648: the first steps toward common, public education in the colonies. After the three laws were passed, Connecticut (1650), New Haven (1655), New York (1665), Plymouth (1671), and Pennsylvania (1683) followed with their own versions.[20]

Adams's claims are difficult to parse. He celebrates liberal education as *national* and liberal while insisting that its origins are Christian and Puritan. Adams's vision of the colonies is—anachronistically—Tocquevillian.[21] Like Tocqueville, Adams maintains that there is less class differentiation and a better educated populace in America. Rogers Smith has noted that this relative equality was true of only a minority of Americans: white, propertied, and male.[22] Smith's criticism of Tocqueville is equally valid for Adams, and Adams's claims are clearly exaggerated. Although Massachusetts *was* the first colony to pass legislation to create local schools, there was resistance in many towns, particularly in areas without Harvard- and Yale-trained ministers. In Massachusetts, there was no formal, tax-supported, state-funded system of schools until the 1840s, and, even then, the schools served barely

50 percent of the population.[23] Moreover, Adams was not entirely committed to universal citizenship as he appeared to reject the education of black Americans.[24]

The statutes passed in Massachusetts did not aim to create a national school system or a national people, yet Adams, writing a decade before the revolution, refers to America. He speaks of a national character prior to the existence of a nation. According to Bailyn, the seventeenth-century colonials were altering educational systems in dramatic ways, but education in the seventeenth century was not about unifying society or national life. Indeed, plans for education and universities that *were* national were rejected.[25] Yet, Adams speaks as if the laws aimed to create a people—a particular kind of liberal people. In sum, Adams characterizes the origins of the system of education as Puritan but insists that the *content* is creedal, liberal, and national. As he discusses the dynamic of rights, he relies on common law principles, but he identifies the origins of public education for liberty-bearing citizens as Puritan.

* * *

Throughout the *Dissertation*, Adams insists that the spirit of liberty is important, but without knowledge, it would be nothing but a "brutal rage."[26] A nation founded on liberal principles requires a populace that is knowledgeable about principles *and* history—particularly the history of rebellions based on rights and the establishment of the American colonies.[27] In order to create an informed populace that ensures liberal government, Adams advocates educational institutions, exercise of rights, and a general enlightenment for freedom. He wants open inquiry: "every sluice of knowledge be opened and set a-flowing."[28] Even as he advocates education as central to this inquiring temperament, he believes enlightenment is the business of all parts of society. Religious leaders should explicitly preach religious liberty. Lawyers should clarify past sacrifices for the establishment of rights and remind citizens that many rights are original, inherent, or essential. Colleges must create public dialogue and disseminate ideas.[29]

Like Brennan and Marshall, Adams insists that public education creates the informed and engaged citizenry upon which American democracy depends. Rights may affect the private life of the individual, but Adams underlines collective benefits: freedom of speech and press embolden citizens to protect liberal government. Because the understanding and exercise of rights is learned, education allows "the people"—particularly the poorer people—to understand and exercise their rights. Adams, repeatedly, invokes the needs of the poor. Even as the poor have fewer resources and less leisure for self-education, they must be educated—at public expense—to control the government and elites. Justice Brennan repeats this theme: education of the children of non-citizens benefits the public as it helps socialize potential citizens.

BOOKS, WITCHES, FOSSILS, AND PLOWS:
THOMAS JEFFERSON'S HOLISTIC SCIENCE

Like John Adams, Thomas Jefferson understood education to ground the critical temperament necessary for popular government. Jefferson embodied the opening of every "sluice" to encourage knowledge and critical discussion. Deeply interested in astronomy, mathematics, meteorology, paleontology, medicine, agriculture, public health, and botany, Jefferson lived his life connecting science and politics. In the same week he became vice president, he delivered a paper on paleontology. He was simultaneously president of the United States and the American Philosophical Society—and much of what we now refer to as science was understood in Jefferson's time to be philosophy or natural philosophy. As a scientist, founder of the University of Virginia, and author of many works dedicated to American schooling, Thomas Jefferson offers a wealth of ideas about education and citizenship.[30] Here, I am concerned with how Jefferson's thought reflects the core elements of the self-governing jurisprudence—public education creates the well-informed public necessary to the maintenance of a democratic regime, and best available science is part of the education that benefits the individual and the nation. Jefferson believed all knowledge to be crucial to the maintenance of democratic government, but he believed science to be particularly important for domestic and global politics.

Jefferson begins by advocating for popular government to protect the rights and freedom of the people. If the people are the "sole depositary of our political and religious freedom," they must guard themselves against tyrants and elites who seek more power. Like Adams, Jefferson did not believe that this collective self-protection of liberty came solely by instinct. In order to resist tyranny and correct the mistakes of their governors, the people needed education—including political history—that would "inform their discretion." Only when the people were well-informed—their minds enlightened, illuminated, or improved—could they vigilantly defend popular government, guaranteeing that liberty was more than a "short-lived possession."[31]

Like Adams, Jefferson did not leave this democratic education to the market. Instead, he insisted that it was the "business of the state" to educate widely: *public* education guaranteed an informed citizenry with the appropriate temperament for self-government. Even as Jefferson spoke of educating the "mass of mankind," he divided citizens into two classes—the laboring and the learned—and recommended education proportional to their "condition and pursuits" in life.[32]

Unless the common people were informed to a "certain degree," popular government would fail.[33] In the interest of the *public*, Jefferson argued for education of the poor—and all citizens—so that they might see their interest in preserving peace and develop the capacities necessary to exercise rights when the order of the government was threatened.[34] Citizens should love their republic—so that they might preserve it—and education would promote that love of the laws.[35] Jefferson went as far as to admire Spain's disenfranchisement of citizens who were unable to read and write.[36] Because Jefferson links ignorance with a lack of freedom, education is crucial to the maintenance of freedom and popular government. Although he valued education for individual liberty and autonomy, he emphasized public education—particularly of the poor—for its collective, national benefits.

Likewise, Jefferson recommended elite education—for the nation—to train honest and capable leaders to "administer the affairs of our country in all its departments, legislative, executive and judiciary, and to bear their proper share in the councils of our national government."[37] Educating for leadership advanced the prosperity, power, and happiness of the nation. Liberal education promoted meritocracy as well as liberty as it encouraged leaders to protect "the rights and liberties of their fellow citizens" rather than allowing those with "wealth, birth or other accidental condition or circumstance" to lead.[38] To this end, Jefferson advocated for elite education for the "youths of genius" from "the classes of the poor." This was consistent with his hope to have an "aristocracy of virtue and talent" rather than one of wealth.[39]

In addition to protecting popular government, education benefited individuals of both classes by developing wisdom and virtue. The ethical development of citizens benefits those individuals as well as the whole. Education is tied to personal development, but virtue simultaneously serves the individual and the community. Wisdom, happiness, virtue, and guardianship of principles are all tied to education—and Jefferson consistently emphasizes the next generation of men who will "guard wisdom" and fulfill duty as well as progressively improving the "condition of man."[40]

In recommending education for both the "laboring and learned," Jefferson emphasized the collective interests of the nation: education creates citizens who will protest tyranny and leaders who will administer honestly and ably to protect the rights of the people. Over many years, Jefferson developed various curricula for primary, secondary, and higher education. In some cases, he explicitly recommended the education of girls in language, math, reading, and history.[41] He was devoted to knowledge as central to progress and the support of free government:

> When I contemplate the immense advances in science and discoveries in the arts which have been made within the period of my life, I look forward with confidence to equal advances by the present generation, and have no doubt they will consequently be ... much wiser than we have been ... than our fathers were, and ... than the burners of witches.[42]

For Jefferson, the "general spread of the light of science" led to "truths"—particularly those that allowed the "mass of mankind" to recognize that they were born with rights such that the favored few could not ride them—legitimately—"booted and spurred."[43] Both the legitimacy and spread of popular government depended upon insights from scientific inquiry.

Science, in the eighteenth century, often referred to general learning—so it is difficult to establish the relationship between education and science in Jefferson's thought. In some cases, Jefferson speaks about science as distinct from the arts or languages, and, regardless of what areas of knowledge Jefferson includes in the term, he establishes that having access to the best available information is crucial to the individual as well as the nation:

> I can assure you, that the possession of it [science] is what (next to an honest heart) will above all things render you dear to your friends, and give you fame and promotion in your own country. When your mind shall be well improved with science, nothing will be necessary to place you in the highest points of view but to pursue the interests of your country, the interests of your friends, and your own interests also with the purest integrity, the most chaste honour.[44]

Jefferson argued that science education was "important to the preservation of our republican government" as well as "essential to its protection against foreign power."[45] He encouraged American political leaders to replicate the European teaching of "every branch of science" to "it's highest degrees."[46]

In his *Memorial on the Book Duty*, Jefferson opposed a book tax on the grounds that the availability of books—particularly those related to science—was essential to scholarship and national security.[47] Jefferson claims that the lack of books in the "different branches of science" and foreign languages unfairly impedes American students in their ability to learn and publish in comparison with their "foreign competitors."[48]

Although Jefferson speaks to the individual learners, he also sees collective benefits. Books, he argues, are not articles of consumption but national capital because they contain "important discoveries and improvements in science and the arts, which are daily advancing the interest and happiness of other nations." Those who study science can be counted on to "become the instructors and benefactors of their fellow-citizens."[49] Thus, scientific knowledge secures liberty and enlightens minds domestically while it protects against foreign powers. Jefferson insists that science is identified with "power, morals, order, and happiness," and Congress should encourage science to that end.[50]

* * *

Like many of his contemporaries, Jefferson thinks about science holistically. He developed techniques for excavating fossils as he was reengineering the plow. He explored the possibility that mastodons were relatives of modern elephants that had adapted to colder climates while he was thinking about yellow fever to advance public health policy. Jefferson understood the study of nature to be inextricably connected to the study of all other disciplines. As such, Jefferson recommended education for a thoughtful citizenry attending, like Adams, to the needs of both the majority and elites. He assumed that having the best available science was in the interest of the individual as well as the nation—and that the national security of the United States was enhanced by access to best available knowledge in the sciences as well as foreign languages. His ideas mirror recent calls for better science education to compete economically in the twenty-first century. Jefferson appears to have seen the start of the nineteenth century in similar terms: a moment in which knowledge drove technology, political principles, personal wisdom, and good government.

GAZING AT THE STARRY HEAVENS: THOMAS PAINE, GOD, AND SCIENTIFIC INQUIRY

Like Adams and Jefferson, Thomas Paine believed that knowledge impacted good government, but he spoke even more directly about science and God. For Paine, a scientific—or questioning—temperament diminishes the ignorance that leads to a lack of liberty. Education created the individual who could be part of the experiment of self-governance. In particular, Paine celebrates how the scientific method's reliance on evidence pushes humans to understand the world without reference to tradition or custom—to see the world as it is (e.g., flat v. round earth).[51] Like Adams and Jefferson, Paine admires how Newton demonstrated *universal* physical principles, and Paine believes using the scientific method will reveal principles in all fields. Because all knowledge is linked, a discovery of the laws governing an eclipse enhances the understanding of political rights. The questioning temperament embedded in the scientific method encouraged a revolution in political principles—popular government—as well as science. Holding humans ignorant of science is the same as holding them ignorant of their rights.[52]

Paine's passion for discovery and the scientific method is accompanied by an enthusiasm for God. If God authors the laws of the universe, scientific inquiry brings humans close to God. As humans gather knowledge through scientific observation, they discover a world created *for* them—by God.[53] Paine not only insists that belief in God is *compatible* with best available science, he concludes that God cannot be properly valued without scientific inquiry. True understanding of God can only be achieved through a scientific understanding of the world. Science is the foundation of theology:

> We can have no idea of his wisdom, but by knowing the order and manner in which it acts. The principles of science lead to this knowledge; for the Creator of man is the Creator of science, and it is through that medium that man can see God, as it were, face to face.[54]

As humans observe the natural world, they gather knowledge about theology and ethics. God has made the earth and "rendered the starry heavens visible, to teach him science and the arts."[55]

Every church is, therefore, a school of science, and God, the "great mechanic of creation," is also the "first philosopher, and original teacher of science." Paine's God invites science and inquiry such that there can be no inconsistency between the structure of "the universe that God has made" and theistic belief.[56] Studying the structure of the universe—without reference to custom or tradition—allows humans to have knowledge of the world and God. We honor God only if we *accurately* map planetary movements or make other observations based on evidence.[57]

Despite the strong connection between God and scientific inquiry, Paine recognizes that his understanding of scientific method is at odds with most organized religion. Paine favored the study of God's creation, but he dismisses the scriptural account from Genesis as the work of men rather than God: "Search not the book called the scripture, which any human hand might make, but the scripture called the Creation."[58] He contrasts scientific study of the divine creation with the "stupid bible" of the church and blames human misery and wickedness on revealed religion.[59] Progress comes from science, not prophets:

> The constant and unwearied observations of our ancestors upon the movements and revolutions of the heavenly bodies, in what are supposed to have been the early ages of the world, have brought this knowledge upon earth. It is not Moses and the prophets, nor Jesus Christ, nor his apostles, that have done it.[60]

For Paine, Christian religions reject evidence in order to promote traditional beliefs that are fraudulent. Paine contrasts revealed religion with the deism he prefers. Christianity is derogatory to God, unedifying to man, repugnant to reason, and self-contradictory.[61]

Paine criticizes Christian churches for persecuting those who use science but clearly believes that the persecution is understandable: revealed religion has much to fear from science. As science encourages a reevaluation of political custom—for example, producing liberal democratic principles—the change in government will be "followed by a revolution in the system of religion."[62] Science, indirectly, will reveal what is false about religion:

> The setters up, therefore, and the advocates of the Christian system of faith, could not but foresee that the continually progressive knowledge that man would gain by the aid of science, of the power and wisdom of God, manifested in the structure of the universe, and in all the works of creation, would militate against, and call into question, the truth of their system of faith; and therefore it became necessary to their purpose to cut learning down to a size less dangerous to their project, and this

they effected by restricting the idea of learning to the dead study of dead languages.[63]

The questioning temperament that popular government requires will be accompanied by an inevitable reevaluation of what Paine calls the fables, myths, tales, or paltry stories that he believes religions—hags of superstition—conjure to fog the minds of humans.[64]

* * *

Paine's writings on science and religion help us better understand the fault lines in today's debates. Whereas supporters of Intelligent Design maintain that evolution implies atheism, Paine insists that a scientific examination of the world leads Americans closer to God. Most world religions—and most versions of Christianity—accept some version of theistic evolution. Like Paine, these religions see God as the author of principles of gravity, astronomy, and biology. Even as Paine's harsh criticisms of the Bible and Christianity contrast with current American convictions, his vocabulary of creation reflects the views of most Americans: God created the world, and evolution was his tool.

Paine maintains that the questioning temperament of science—with its insistence on evidence—threatens any religion that maintains that scripture is literal. Literalist faiths that require the words of Genesis to be the only account of God's creation of the physical world and, more importantly, human beings *should* be threatened by the teaching of evolution and the questioning temperament that Paine celebrates.

EARTHQUAKES, ECLIPSES, AND EVOLUTION: CREATING CIVIC TEMPERAMENT THROUGH EDUCATION

Like Brennan, Marshall—and all the other justices who have written on education over the last 90 years—Adams, Jefferson, and Paine assume that education is necessary for the maintenance of democratic government. First, rights may not be exercised against threats to liberty unless citizens are educated to understand these rights. Here, Adams and, to a lesser extent, Jefferson make arguments that mirror those made by Thurgood Marshall. In order for freedom of the press and speech to prevent power grabs by potential tyrants or power-hungry elites, the people must be educated. Education is linked to specific rights and a general ability to think and discern. Second, all agree that the education of the poor must be the business of government. Whereas Brennan and Marshall speak to both the collective and individual benefits of public education of the children of immigrants, Adams and Jefferson tend to emphasize the effect of educated citizens on the safety of the nation. Yet all underline the need for education for the good of the nation.

Unlike the justices of the twentieth century, Jefferson and Adams explicitly call for education of elites to create leaders for all branches of government.

How does *science* matter to this political education? Adams, Jefferson, and Paine believe that a questioning temperament is crucial to the establishment and maintenance of democratic citizenship—and they believe that science is part of that mindset. As college-educated elites, they had been educated in scientific thought, and they integrated the physical and biological sciences—as well as classical languages and math—into their understanding of "enlightenment."[65] This holistic way of thinking about education is reflected in the vocabulary of their time—in which science and philosophy had expansive meanings that included the hard sciences as well as classical drama or theology—as well as the way they wrote about politics. Science and politics cross-pollinate. In explaining bicameralism, John Adams referred to Isaac Newton's *Principia* because he believed political—the balance of powers—and physical equilibrium to be related or the vocabulary of science clarified his political point.[66] Science is one of the foundations for critical inquiry, and Jefferson and Paine underline the role of science in reevaluating custom and legitimizing popular government.

This holistic way of thinking about inquiry, science, and politics also extends to God. Paine sees no contradiction in God creating the universe and humans interrogating that creation. Adams can believe that God shook the earth during the Lisbon earthquake yet reject Reverend Prince's claim that God punishes scientists like Benjamin Franklin for inventing lightening rods.[67] Like Paine, Adams believes in a God that creates natural laws and forces—inviting humans to study them. Understanding the natural order does not, inevitably, lead to atheism. John Winthrop—a Puritan, not a deist—dives into the debate over earthquake science without fear that he will be called an atheist. Yet Paine helps us understand the fear of science education among literalist religions. If science creates an inquiring spirit that leads humans to question political traditions, it also encourages reevaluation of religious custom and scripture. Science does not threaten all believers— but it does challenge key beliefs of literalists (e.g. young earth creationism).

Some scientific organizations have tried to provide the equivalent of Winthrop's lectures and pamphlet. As they defend the science of evolution, they explicitly state that belief in evolution is consistent with religious faith, many scientists are believers, and the works of science are not intended as a commentary on matters of faith. Given Paine's insight on religion and the critical spirit, any successful approach should steer clear of science versus religion and explicitly defend against the claim of ID that belief in evolution is tantamount to rejection of God.

As we try to build a new vocabulary for discussing the role of science in American political life, there are several important lessons.

First, American liberalism requires a knowledgeable citizenry and science is part of that knowledge, but the critical spirit that Adams, Jefferson, and Paine recommend—in which "every sluice of knowledge"[68] should be opened in all arenas of life—may be more than Americans currently accept. Although the American political tradition and constitutional structure depends upon the people to be vigilant and knowledgeable, we are in a political moment in which education is something of a liability: to be invested in reason or processes of science and logic is suspect. In celebrating the 250th anniversary of Jefferson's birth, lectures focused on Jefferson's interests in religious freedom, classics, and architecture—but few were directed at his scientific work. Americans appear uncomfortable with the role of science in the founding of the nation.[69]

At the time of the founding, being educated—especially in science—was not a liability, and the vocabulary of self-governance must make room for science that does not imply atheism. It seems necessary to remind Americans of the tradition of critical thinking—framing it as patriotic as well as godly—and encourage them to reinvest in this vision of American democracy.

6 Religion and Science
Refuting the Conflict Narrative

Reading the works of Adams, Jefferson, and Paine leads us to reflect on the remarkable differences between the political narratives of today and the eighteenth century. One of the most vivid differences is the lack of a conflict between religion and politics. Winthrop—a man devoted to God as a minister—comfortably takes up the claims of another minister regarding the science of earthquakes. In the eighteenth century, Paine assumes that studying the stars is the same as knowing God. In the twenty-first century, Representative Paul Broun Republican-GA insists that evolution, embryology, and the Big Bang theory are "lies straight from the pit of hell." Broun stridently declares that the earth is 9,000 years old, even though scientists have dated the oldest human fossil at 195,000 years old. Broun is a physician and member of the House Science, Space, and Technology Committee.[1] How did we move from an integration of God and best available science to the polarized rhetoric of today's politics? Some answers lie in how Intelligent Design and evolution advocates describe the relationship between science and religion. But a more careful analysis of how religions approached Darwin in the nineteenth and twentieth centuries demonstrates that this narrative should be rejected.

As we saw in Chapter 1, Intelligent Design depends on two closely related claims: teaching evolution falsifies religion, and evolution necessarily implies atheism. Together, these claims help ID create a powerful narrative: the *inevitable* conflict between evolution and religion. For ID, the conflict narrative motivates citizens to complement evolution (atheism) with a more God-friendly approach (ID) in the public school science curriculum. ID has an unlikely ally here as advocates of evolution who insist upon atheism employ the same narrative of incompatibility. Evolutionary biologist Richard Dawkins maintains that evolution reveals the farcical nature of religious belief and implies and encourages atheism.[2] Despite the claims of ID and Dawkins, this inevitable conflict model is, according to most major religions and most scientists, false.

Two minorities—literalist Christians and devote atheists—insist on the inevitable conflict, but most of the West's religions remain firmly and consistently committed to a narrative of *accommodation*. In the twenty-first century,

Roman Catholicism, most forms of Judaism, many Muslims, and most denominations of Protestantism accept *both* evolution and belief in a monotheistic God by appealing to some version of theistic evolution: the world was created by God, but humans evolved according to natural laws. Literalist Christians see an inevitable conflict between Genesis and evolution, but most Christians do not employ literalism as the principle for governing the exegesis of sacred texts.[3] Intelligent Design and the inevitable conflict model not only corrupt the history of religion and science for its own political purposes, but also ignore the fluidity of both terms over time and within each religious tradition. In the West, religion and science have overlapped, confused, complemented, and conflicted. Each religion frames the challenge of science differently—and within each religious tradition there is diversity of belief. For example, the Roman Catholic Church reacted to sun-centered astronomy, but Islam did not. "Religion" does not react to "science." Rather, particular scientific ideas challenge specific religions or beliefs within those religions. Despite their differences, Judaism, Islam, and Christianity all understand God to be the author of two books: a sacred book of scripture and the book of nature. To differing degrees, these religions maintain that their faithful should read *both* books. Within each faith tradition, there is debate over the relative authority of different sources of knowledge. All three religions believe in revealed knowledge (through scripture or directly to individuals by God) as well as knowledge obtained by humans through their reason and interaction with the material world.[4]

This chapter provides a brief overview of how major religions reacted to Darwin's theory of evolution in the nineteenth century as well as how they have adapted their views over time to demonstrate that Western religions have, in general, accommodated evolution. In the context of the United States, Intelligent Design disingenuously represents Christianity's relationship with evolution. ID's narrative—of conflict and literalism—should not be adopted by courts or political theorists. Liberals should insist—as most religions do—that the theory of evolution does not require rejection of God or religious practice.

JUDAISM AND ISLAM

For most of the world's Jews, evolution is consistent with and substantiates their interpretation of sacred texts and the rabbinic tradition: Torah,

Midrash, and (for some) Kabbalah. The Jewish paradigm of compatibility dates from the medieval period. Although individual Jews played an active role in scientific culture, particularly in the Islamic world, medieval Judaism never had a significant tradition of natural theology: scientific knowledge was not sought to establish the existence or attributes of God. For example, the twelfth-century physician and philosopher Maimonides saw no conflict between revelation and reason. Maimonides attributed any contradictions between scripture and human observation to humans' imperfect understanding of God, God's world, or both. For Maimonides, study of the natural world was both an autonomous pursuit *and* a path toward a deeper understanding of God.[5]

In the nineteenth century, most Jews did not react to *On the Origin of Species* as threatening to religious belief. Chemist Raphael Meldola (1849–1915) corresponded with Darwin and defended natural selection. Scholar of Talmud and Torah, Naphtali Levy, wrote to Darwin (in Hebrew) insisting that the dust of Genesis might symbolize the long process of evolution from inorganic matter to the emergence of humankind. Most nineteenth-century rabbis maintained that God created the universe and evolution operated as a natural process within that framework: theistic evolution. Some rabbis focused on the consistency between evolution and the Kabbalah's narrative of change.[6]

Although modern Jews—reform, conservative, orthodox, haredi (a conservative form of orthodox), and Hasidic—differ in their understanding of *how* the Torah is consistent with evolution, most find no contradiction between evolution and fundamental Jewish belief. With few exceptions, modern Jews accept that the earth is billions of years old. They differ on how to derive this number. While some understand each day of Genesis as an epoch rather than a conventional day, others use Midrash to claim that God created previous worlds before deciding on the current one. Even as most denominations accept evolution alongside some form of theistic evolution, they disagree about the extent to which God intervenes (e.g., God actively promotes genetic mutations or transmutations of species over time, or natural processes alone guide evolutionary change).[7] Despite the general agreement, orthodox and Hasidic traditions tend to focus on Torah rather than the study of science (or other secular subjects), and some Haredi and Hasidic rabbis have actively opposed evolution. The pro-evolution works of orthodox Rabbi Natan Slifkin, for example, were banned by a group of U.S. and Israeli orthodox rabbis in 2005. Slifkin promotes the study of natural history as compatible with traditional Jewish texts, arguing that the "days" of Genesis are six long eras, each billions of years long.[8] In sum, most of the world's Jews accept evolution as compatible with Jewish scripture and belief. Despite some differences, all Jewish denominations are united in their opposition to teaching Intelligent Design in public schools as the injection of Christianity into the public sphere.

Like Judaism, modern Islam embraces varying forms of theistic evolution. However, an active and possibly growing force within modern Islam actively opposes evolution as hostile to fundamental Islamic beliefs and scripture. Although Muslims are not the focus in public school debates (and American polling data does not capture their convictions fully), their religious beliefs may increasingly affect the debate in areas where Muslims make up larger percentages of public school populations.

Similar to the Genesis narrative, the Qur'an teaches that Allah created the seven heavens and the earth in six *ayums*, crafted humans out of clay, and created each species separately. Because *ayums* might be stages or days, young earth creationism is not an issue for most Muslims.[9] Many Muslims present evolution—and scientific discovery generally—as consistent with medieval Islam's rich intellectual tradition of natural classification and support of separate spheres of knowledge for science and religion.[10] From the late eighth to the sixteenth century, Islamic cultures were engaged in scientific discovery, including astronomy, mathematics, optics, and medicine.[11] Well before Darwin, al-Jahiz (776–869) described resource competition among organisms, and Nasir al-Din al-Tusi (1201–1274), a contemporary of Thomas Aquinas, connected new features of animals and plants to survival advantage in their environments.[12]

Muslim harmony regarding science and faith continued until the last quarter of the nineteenth century when influences from the West triggered some opposition to science as materialism. The earliest traceable mention of Darwin's theory of evolution appeared in 1879, but the first five chapters of *On the Origin of Species* were not available in Arabic until the 1920s.[13] Because of national ties with France, many late nineteenth-century Muslim scholars focused more on Lamarck than Darwin. When Darwin was debated, the focus was often on *social* Darwinism, and the reaction was negative.[14] In general, traditionalists, such as members of the old *ulama* class, accommodated evolution, while intellectuals educated in more urban settings used science to oppose religion.[15]

These Islamic responses to Darwin reflect direct and indirect influences from the West. Some Islamic scholars presented Islamic study of religion and science as mutually reaffirming by opposing Western texts such as John William Draper's *Conflict between Religion and Science* (1874).[16] But financial ties between Islamic universities and American Protestant churches blurred some debates. The separation of religion and science by Muslim scholars often was aimed at the American Protestant patrons rather than Islamic scripture.[17] In *Refutation of the Materialists* (1881), Jamal-al-Din al-Afghani (1838–1897) opposed Darwin's theory of evolution as part of Western materialism and a denial of the existence of God. Later, he supported evolution to explain the development of animals, but he saw humans as special creations given their souls.[18] Although Shibli Shumayyil (1850–1917), a graduate of the Syrian Protestant College, also criticized materialism, he strongly supported evolution and argued for a separation of science and faith.[19] Hussein al-Jisr's (1845–1909) approach was far more common. He found natural selection to be compatible with Muslim cosmology and faith.[20]

Contemporary Islam includes theistic evolution, rejection of modern science, and promoters of creationism and Intelligent Design, such as Adnan Oktar, who cast evolution as atheistic, materialistic, immoral, and threatening to Islam's rejection of random events (e.g., genetic mutations).[21] Other Muslims accommodate science by, for example, searching the Qur'an for passages that foreshadow modern scientific discovery in order to prove that the Qur'an—written fourteen centuries earlier—is miraculous because of its scientific prescience.[22]

CHRISTIANITY

As American courts try to determine whether Intelligent Design is excludable religion, the focus has been on the history of Christian fundamentalism.[23] With over 1,000 sects of Protestantism in the United States, it is impossible to summarize the views of all groups. My purpose here is to establish that the Christian tradition has a rich history of accommodating evolution. I begin with some background on Catholicism—the largest denomination in the United States—and then I discuss some aspects of Protestantism.

Although often identified with the seventeenth-century trial of Galileo for his support of heliocentrism, the Roman Catholic Church supported and funded scientific research throughout the Renaissance. The Jesuits received generous funds for their astronomical observatories, and the Roman Catholic Church gave "more financial and social support to the study of astronomy over six centuries, from the recovery of ancient learning during the late Middle Ages into the Enlightenment, than any other, and probably, all other, institutions."[24] Although the sixteenth-century Church insisted that heliocentrism contradicted scripture, earlier Catholics had maintained compatibility between the study of nature and scripture. For Augustine, humans acquired knowledge through reason and experience of the "motion and orbit of the stars" and "predictable eclipses of the sun and moon." If Christians ignored this knowledge of the material world, Augustine cautioned, they would place Christians in "an embarrassing situation, in which people show up vast ignorance in a Christian and laugh it to scorn." Ignorance of science would make Christianity less plausible to non-Christians they were trying to convert.[25]

Catholicism's long tradition of natural philosophy encouraged the integration of science and faith, including Pierre Gassendi (1592–1655), René Descartes (1596–1650), Blaise Pascal (1623–1662), and Nicholaus Steno (1638–1686).[26] Writing to the Catholic theologian Marin Mersenne (1588–1648), Descartes insisted that "God sets up mathematical laws in nature as a king sets up laws in his kingdom."[27] Many of these early Catholic thinkers assumed—as Newton and Boyle would—that God had designed the universe with universal, constant, and mathematical laws that humans could discover rather than governing through constant intervention, even though they assumed that God had that power.[28] Despite finding Galileo guilty of heresy and some insistence on a literal reading of some scripture, the Roman Catholic Church accommodated most science and supported using reason to explore a world that God had designed.

By the nineteenth century, the Church had moved away from literalism, favoring a more symbolic reading of Genesis that extended the age of the earth (making it more consistent with nineteenth-century findings in geology). Voices within the Church, such as the nineteenth-century anatomist St George Mivart, tried to persuade Rome of the plausibility of theistic evolution.[29] Darwin's *On the Origin of Species* elicited no authoritative response from the Church. The book was neither seen as a work of atheism nor placed on the *Index Librorum Prohibitorum* (the list of books prohibited by the Church to protect the morals and faith of adherents). Response to evolution was varied, but many in the Church—like John Henry Newman—believed that God created the laws of nature to be discovered by humans:

> Mr. Darwin's theory need not then to be atheistical, be it true or not; it may simply be suggesting a larger idea of Divine Prescience and Skill. Perhaps your friend has got a surer clue to guide him than I have, who have never studied the question, and I do not [see] that 'the accidental evolution of organic beings' is inconsistent with divine design—It is accidental to us, not to God.[30]

The Church allowed that Darwinian processes explained the physical evolution of plants and animals but insisted that *human* evolution was exceptional. First, each individual human soul is created in the image of God and cannot be explained merely as the product of materialistic evolution (special transformation). Secondly, all humans descended from one pair of ancestors: Adam and Eve (monogenism). At the turn of the nineteenth century, the Church opposed scholarship that did not endorse special transformation or argued for polygenism (multiple first men and first women as human ancestors) by placing it on the *Index*.[31]

By the middle of the twentieth century, the Jesuit paleontologist Pierre Teilhard de Chardin interpreted evolution as a divinely guided process with human moral and spiritual awareness as its goal,[32] and Pope Pius XII issued the encyclical *Humani Generis* (1950), announcing that Catholics had academic freedom to research evolution as long as Catholic dogma was not violated. In particular, polygenism must be rejected, and the soul must be understood as created by God. By the late 1990s, Pope John Paul II reaffirmed special transformation but emphasized that evolution was now more than a hypothesis. The modern Roman Catholic Church does not require a literal understanding of Genesis or belief in a young earth. For Roman Catholics, evolution is a natural process that is planned and guided by God.

Their theistic evolution allows for most scientific findings but insists that God guides evolution and is solely responsible for the soul. All individuals have purpose outside of evolution's natural forces.

The papacy of Benedict XVI created confusion over the Church's position on both evolution and Intelligent Design. Speaking at his inaugural mass in 2005, Pope Benedict XVI accepted evolution as science but forcefully reaffirmed that humans are purposefully created by God and different from animals and plants: "We are not some casual and meaningless product of evolution. Each of us is the result of a thought of God. Each of us is willed, each of us is loved, each of us is necessary." Evolution, the Pope warned, cannot be a worldview that deprives the world of meaning and purpose.[33] Although the Roman Catholic Church has not generally been supportive of Intelligent Design, Benedict XVI and some of his close advisers appeared to both reject and endorse some version of design. Despite these statements, there has been no official change in Church doctrine since *Humani Generis* (1950), and prominent institutions associated with the Church have continued to support theistic evolution. In 2004, an International Theological Commission accepted the big bang, genetic relationship of organisms, theory of evolution, and the age of the earth as 15+ billion years old. In 2009, a conference of Catholic scholars affirmed the lack of conflict between evolutionary theory and Catholic theology and emphatically rejected ID. Although Pope Francis is trained as a chemist, little is yet known about how he will lead the Church with regards to evolution.

While it is impossible to summarize the views of Protestant Christianity—with its multiple denominations—a brief review of the reaction to Darwin in England and the United States demonstrates that most Protestants accommodate the language of the scriptures with new scientific discovery.

For Protestant churches, *On the Origin of Species* followed 50 years of scientists and clergymen reconciling new evidence in the fossil record and geology with the scriptural six days of creation.[34] The publication of Darwin's work also coincided with a wide-ranging debate in the Anglican church over how to interpret the Bible. Samuel Wilberforce—who would later famously challenge Darwin—was more disturbed by essays that argued for reading the Bible as any other book than the theory of evolution.[35] Darwin did not date the earth or explain the origins of the universe. Darwin's claim—that natural selection served as the mechanism of change and adaptation to an environment—challenged the notion of a designing, deliberate creator who created humans in God's own image. Some Anglicans—like Catholics in the nineteenth century—worried about the status of the soul and the extent to which humans were God's special creation.[36]

The 1860 debate between Bishop Samuel Wilberforce and Thomas H. Huxley at the British Association for the Advancement of Science is often cited as the paradigmatic representation of the conflict narrative. In his review of *On the Origin of Species*, Huxley had claimed that evolution killed religion: "Extinguished theologians lie about the cradle of every science as the strangled snakes beside that of Hercules; and history records that whenever science and orthodoxy have been fairly opposed, the latter has been forced to retire from the lists, bleeding and crushed if not annihilated; scotched, if not slain."[37] During this debate over *On the Origin of Species*, Wilberforce asked Huxley whether he claimed maternal or paternal descent from a monkey, and Huxley is reputed to have replied, "he would not be ashamed to have a monkey for his ancestor, but he would be ashamed to be connected with a man who used his great gifts to obscure the truth."[38] Interestingly, Wilberforce objected to Darwin based on contradicting evidence from contemporary works of *science*. Darwin had claimed that cruelty and imperfections in the natural world—particularly the larval cannibalism ichneumon wasps—might be more easily explained by evolution than a creator who allowed cruelty.[39] Wilberforce countered that all of nature fell from God's grace into disorder when Adam and Eve were expelled from the garden. Wilberforce did not endorse biblical literalism, but he championed human exceptionalism. Evolution demeaned God and humanity if humans were not subject to a special creation.[40]

Some English Christians found Darwin's natural theology an appealing means by which to conceive of God as the creator of law-like processes. In Scotland, Henry Drummond (1851–1897) saw ways to combine natural law and spirituality. In a series of lectures given in Boston in 1893, he argued that a slow evolution was just as miraculous as a six-day process. God's evolution had produced mountains and valleys as well as love, "the final result of evolution." Drummond's theology associated God with human knowledge rather than ignorance. He rejected the notion of a "God of the gaps" and instead saw God in the general process of evolution.[41] Drummond's approach seems relevant today as ID's Designer explains what is *outside* of the laws of nature rather than what is inside. The future Archbishop of Canterbury, Frederick Temple, preferred a God of natural law, variation, and natural selection to one that altered the world by quick miracles. In the United States, the Harvard botanist and Presbyterian Asa Gray (perhaps the most important and consistent supporter of Darwin in North America) also adopted a theistic version of evolution: natural selection under divine control.[42] The Anglican church honored Darwin—as they had Newton—with a funeral at Westminster Abbey. Comparing Darwin with Newton, Reverend Frederic Farrar publically maintained that the theory of evolution by means of natural selection was consistent with God as the creator of the natural world.[43]

Modern Anglicans describe their faith as a "three-legged stool" of Scripture, Tradition, and Reason and insist that this has been their position since the sixteenth century. Specifically, they claim that scientific exploration of the origins of life does not conflict with the "central truths of Scripture and Christian Tradition."[44] In the United States, Episcopalians point to priest-scientists who have shown that "an evolutionary world view can be integrated with a theology of creation."[45] In general, the Episcopal Church has steadfastly supported the harmony of science and Christian belief. In 1982, they passed a resolution contending that "the glorious ability of God to create in any manner" did not conflict with the search for truth by scientists in that creation. They explicitly rejected the "rigid dogmatism of the Creationist movement."[46] In 2006, at the time of the Dover controversy, the Episcopal Church emphatically rejected any contradiction between the Christian belief in God as the creator and the theory of evolution as providing a "fruitful and unifying scientific explanation for the emergence of life on earth." Evolution does not diminish the centrality of scripture in "telling the stories of the love of God for the Creation and [is] *entirely compatible* with an authentic and living Christian faith." The resolution explicitly encourages state legislatures and local school boards to teach best available science.[47]

In the nineteenth century, Christians used Darwin to support theories of race or defend human equality. Some Christians used the Bible to support monogenism (all humans derive from Adam and Eve) and a unified humanity based on equality. Others employed polygenism (humans derive from different ancestral parents) to justify slavery and racial inferiority. Polygenism threatened Catholic *and* Protestant theology's commitment to special creation. Darwin's version of monogenism—found in *The Descent of Man* (1871)—was not like earlier Christian versions (that insisted that equality followed from monogenic descent) because Darwin allowed for changes in different groups based on their environments and random mutation. Darwin upset the most insidious versions of polygenism (belief in separate and permanent differences), but he allowed for continuous selection that might produce different races, and his ideas were later used by the eugenics movement: a controlled breeding for "desirable" traits. Darwin was used to justify, for example, the removal of indigenous peoples in New Zealand and racial eugenics in the late nineteenth and early twentieth century. In Britain, economist Ronald Fisher (1890–1962) believed it was Christian duty to breed for "higher ability, richer health, and greater beauty."[48] During the early twentieth century, eugenics had the support of other scientists (e.g., Madison Grant and Robert Yerkes) and public intellectuals (e.g., Winston Churchill, Theodore Roosevelt, and Margaret Sanger). By the late 1920s, eugenics was part of mainstream scientific culture in the United States with 376 American colleges offering courses on eugenics. The presidents of Yale, Harvard, Stanford, the University of Michigan, the University of Wisconsin, and the University of California, as well as the head of the American Museum of Natural History—all supported eugenics. The science of eugenics was used to oppose South and Eastern European immigration and marriage (e.g., Immigration Act of 1924).[49] At the Scopes trial, William Jennings Bryan—looking particularly at *The Descent of*

Man—linked racial engineering and eliminating the poor and weak from society with the teaching of evolution.[50]

CONCLUSION

This brief overview can help us better understand how the incompatibility narrative—science versus religion—does not fully represent the beliefs of all citizens—and why science for citizenship should look to recapture an account of science and religion as compatible and, sometimes, complementary.

When Darwin published *On the Origin of Species*, there was *not* unified religious condemnation. Despite the remarkable pluralism among and within religions, Jews, Muslims, and Christians applied ideas about faith—developed over centuries—to Darwin's suggestion that humans evolved over time. Most developed some version of theistic evolution. Earlier in the century, many of these religions had revised their claims about the age of the world as new fossil and geologic evidence emerged. In contrast to today's young earth creationists, most Christians have long assumed that the language of the scriptures was designed to accommodate the limitations of understanding of its original readers rather than conveying scientific information about the earth.[51] As courts debate Intelligent Design, they explore the rise of religious opposition to evolution in the American case and, therefore, focus on the rise of Christian fundamentalism and its attachment to anti-evolution. Judges legitimately look to this history to establish historical context and the *intent* to inject religion into the public schools.[52] As we focus on these cases and listen to the rhetoric of Intelligent Design, we may miss an opportunity to use the wider *agreement* about theistic evolution to gain support for evolution education. What is required is a truer narrative of the separation of church and state: an insistence that science does not oppose religion and, equally important, that science does not settle religious questions.

Attempts to respect religion and establish another narrative have often been met with scorn by evolution advocates such as Dawkins. In an effort to make evolution education more compatible with religion, the National Association of Biology Teachers (NABT) voted to delete the words "unsupervised" and "impersonal" from a statement regarding evolution: "To say that evolution is unsupervised is to make a theological statement," and that exceeds the bounds of science.[53] Responding to this attempt to separate claims about God from good science teaching, Dawkins accused the group of "cowardly flabbiness of the intellect."[54] The opposition narrative works for both ID and evolution—even as they appeal to the clash between Darwinism and Christianity—for different reasons. Democracies should, in the public square, encourage this clash of worldviews, but the NABT approach makes it possible, in the public schools, to teach best available science without a debate about religion.[55]

This historical overview helps demonstrate why the NABT approach furthers science for citizenship while preserving the separation of church and state. The conflict narrative disingenuously empowers Christian literalists to legislate "for God" by opposing evolution. A commitment to religious accommodation (the theistic evolution accepted by Roman Catholics, most Protestant denominations, and most forms of Judaism) leads to a truer separation of church and state. Science does not take a stand on whether the universe is supervised by God and religious belief is not the justification for what is taught as science. The conflict narrative has distorted efforts to deliberate on the content of science curriculum. Literalists *do* believe that teaching evolution denies biblical truth but the West's faith traditions are far more heterogeneous and complex. The narrative of accommodation—like NABT's new language—yields science education policy that is true to most religious faiths without compromising American constitutional principles.

Conclusion
Making the Case for Science in Liberal Democracy

When the Soviet Union launched the Sputnik satellite in 1957, the United States quickly moved to improve science education and research—as essential to national security and global economic competition. By 1958, President Eisenhower had directed billions of dollars into education through the National Defense Education Act and created the National Aeronautic and Space Act: NASA. Later, President Kennedy encouraged Americans—as part of their commitment to the nation—to study and advance scientific knowledge. Addressing the National Academy of Sciences in 1963, Kennedy argued that "science contributes to our culture in many ways, as a creative intellectual activity in its own right, as the light which has served to illuminate man's place in the universe, and as the source of understanding of man's own nature."[1] Fifty years after Sputnik, a majority of the Republican candidates for the presidency rejected evolution and affirmed Intelligent Design—relying on the language of critical inquiry to establish scientific "controversy." Republican Rick Perry told a young boy that Texas teaches both creationism and evolution because "you're smart enough to figure out which one is right," and Michele Bachman supported Intelligent Design because government should not takes sides on scientific issues when there is "reasonable doubt on both sides." Breaking with fellow Republicans, Jon Hunstman tweeted, "I believe in evolution and trust scientists on global warming. Call me crazy."[2]

Perry and Bachman's language of choice and open discussion is now a political weapon—used to support teaching creationism or discredited science like Intelligent Design. We see this logic embedded in the new "critical inquiry" laws in Tennessee and Louisiana: declare the existence of a scientifi c controversy without any scientifi c evidence, and require students to study it. Those who support ID use the language of democratic deliberation—expressing diverse ideas, teaching controversies, evaluating strengths and weaknesses, and supporting academic freedom—in order to present faith-based understandings of human origins.

Debates over Intelligent Design, creationism, and critical inquiry begin with confl ict over religious belief, but, ultimately, they expose a fundamental weakness in American political and legal thought: the inability to articulate the connection between science education and democratic practice. Informed by contemporary theory, court decisions, and eighteenth-century American works, this book urges that we explore the relationship between science education and democratic citizenship. Rather than defensively excluding Intelligent Design as religion, liberals should argue for the inclusion of best available science. Framing the debate in terms of education for a democratic temperament allows liberals to argue that democratic politics depends upon education that fosters critical thinking in order to nurture citizens—and science is part of that education. This new narrative allows us to take back the vocabulary of democratic practice as well as to build an alternative jurisprudence—based on education precedents—that can effectively protect science education from "critical inquiry" laws. Our discussion of science education should center on citizenship, information, and public reason.

Why the focus on science? Why isn't it sufficient to exclude religion? Because courts use the Establishment Clause to remove creation-science or Intelligent Design from science curriculums, it is difficult to cut through to issues of science and democratic citizenship. Like earlier courts, the Dover majority excluded Intelligent Design from the science curriculum as religion. Yet this Establishment Clause approach has two serious weaknesses.

Courts rely upon the separation of church and state to determine whether teaching religious doctrines in the public schools favors some students over others. But the Establishment Clause jurisprudence only works when the material is religious. Because new laws claim to enhance "critical inquiry" by introducing "all views" into the science classroom on "controversial" scientific subjects, courts would be forced to demonstrate that the alternate approaches are religious. This gets more difficult as Intelligent Design wraps claims in the language of science or creates journals for peer review. Excluding ID as religion does not address whether discredited science can be taught. The Establishment clause strategy must exclude material as religion rather than defend teaching best available science for democratic purposes. The legal vocabulary of this jurisprudence has practical political consequences as well. Citizens appear to believe that when courts *exclude* religion in a decision, they are *opposing* religion. They then link teaching evolution with supporting atheism.

The rise of "critical inquiry" laws should not come as a surprise because a majority—between 56–65 percent—of Americans have told pollsters that they want *both* creationism and evolution taught in public schools. Over the last three decades, courts have ruled against the will of the majority.[3] The constitutional system depends upon courts to uphold unpopular rights or requirements, but it would be unwise to ignore a growing democratic problem: scientists believe there is a consensus that the majority does not recognize. The problem is exacerbated when scientists such as Richard Dawkins dismiss the beliefs of the average citizen as stupid or delusional.[4]

While courts can continue to protect equal citizenship using the Establishment Clause, new legal and political changes may also require explaining why teaching scientific controversies using discredited science offends the Constitution. Politically, it is equally necessary to address the perception that excluding religion means opposing God. In an effort to address both the legal and political issues, this book recommends a focus on basic skills for citizenship and the development of a democratic temperament. Given an American citizenry that self-identifies as religious and democratic simultaneously, focusing on science for citizenship—rather than autonomy or liberal values—offers a principled and politically practical approach. The ideas of John Adams and Thomas Jefferson—education to create tyranny-preventing citizens who speak and publish—complement the logic of the *Plyler* decision with its call to educate students for political and economic life. Works in political theory help flesh out the type of skills necessary for the democratic temperament. Together, these ideas create a better defense against the new critical inquiry laws or the new political language of "inclusion."

Importantly, science for citizenship is neither antagonistic to religion nor elitist. During the Scopes trial, H.L. Mencken referred to creationism as "degraded nonsense which country preachers are ramming and hammering into yokel skulls."[5] In the twenty-first century, biologist Richard Dawkins combines important data to discredit ID with his claim that faith is a delusion. He includes *ad hominem* attacks on those who don't believe in evolution as stupid or insane.[6] Given the complexity of American religious culture, science for citizenship insists that science in public schools limit its inquiry to the natural world—leaving questions of spirituality and belief to individuals. This is not only a more principled approach to the separation of church and state, it is also the position of most of the West's religiously faithful; most denominations have—since the publication of *The Origin of Species*—found ways to reconcile the theory of evolution with their beliefs and practices. Thomas Paine's deism does not work in twenty-first-century America, but, according to public opinion polls, more Americans accept evolution if theistic evolution is included: (54% in 2010; 47% in 2012). Insisting—as Dawkins does—that evolution requires atheism exacerbates battles over science education as only 15 percent of Americans accept evolution without any role for God.[7] If evolution must mean atheism, the state risks increasing the number of citizens who believe creationism and evolution should be taught in public schools (currently between 56% and 65%). Attending to these facts of American pluralism should lead liberals to insist that science investigates the natural world and avoid unnecessary attacks on religious believers as "insane." This approach helps bolster the claim that science education is not about religion, and evolution is not (as many creationists insist) a faith that is being taught in schools. The National Academy of Sciences has moved in this direction—carefully wording their publications to clarify that belief in evolution does not require a choice between God and science.[8] The science community will have more support from the majority if we can return to John Adams and the earthquake: a time when a minister like Winthrop could speak and publish about the science of earthquakes without fear that he would be considered godless.

Looking at works by Adams, Jefferson, and Paine, as well as the many ways that Western religions reacted to Darwin in the nineteenth and twentieth centuries, helps us shape this type of approach. Since Aristotle, Western thinkers have grappled with a framework for knowledge: how and what should humans study to understand their world? What do they need to learn to be citizens or to act ethically? What we now call science has been a part of that understanding, but the emergence of a scientific culture took many centuries, and its relationship with religion has been extremely fluid. In modernity, we have the model of Descartes (God is inscrutable, and we cannot learn anything about God from natural philosophy) as well as Boyle and Paine (the study of nature yields a better understanding of God).[9] Intelligent Design capitalizes—and fuels—a narrative of opposition: science versus god. Science for citizenship must reject this narrative by appealing to the history of how religions have reconciled the findings of science, especially evolution, with their spiritual foundations and beliefs. The relationship between science, religion, and politics is not fixed.

Thinking about science education and political citizenship also reveals a weakness in liberal political thought. Legitimate concerns about positivism—representing science as a neutral standpoint—often distract liberals from engaging with crucial issues of fair play, science, and education. Justice Scalia, ID supporters, and the Tennessee and Louisiana critical inquiry laws present a variation on postmodernism: all theories are equally valid because there exist no criteria to determine truth. To be considered, one just needs a few supporters who are willing to make the argument. Scalia insisted that the evidence for creation-science was just as strong as evolution, even though evolution has the support of 99.85 percent of life scientists. Liberals should be clear that disagreement does not rule out making judgments, and public reason depends upon our ability to argue based on evidence. A liberal political education cannot fetishize science as an objective arbiter of truth, but disagreement about evolution does not mean that all arguments are equal. It is possible—and essential—to judge some arguments better than others by judging which ones are more "plausible, persuasive, or convincing."[10]

Science is not the only mechanism for making such judgments, but science helps to train citizens to deliberate and acquire and evaluate information. As such, the United States must include scientific method and facts about the natural world as part of the education that encourages a democratic inquiry. Solid judgments on climate change or bio-engineering, for example, require substantive knowledge of climate science and biology.[11] As future citizens, students need to know what scientists have determined—by experimentation over time—is the best possible interpretation of the evidence. Liberals must articulate positive arguments for teaching science and explore how science undergirds secular standards of reasoning. The debate over ID helps expose democratic dependence on public reason—and the need to support public reason through education.

Making these claims requires careful attention to free speech in the context of educating minors. In the public square, individuals and groups are free to make any claim they like about human origins—and the number of Internet sites dedicated to creationism and Intelligent Design speaks to how established that freedom is. Although schools are important sites for free speech, courts have always limited speech in the name of education. One essential limit must be teaching information that has been discredited as science—information that is *false*—as well as suggesting that all opinions are equally valid. These two requirements of "balanced" or "critical" education negatively affect the students' capacity to weigh evidence and develop critical inquiry skills.

The issues raised by my analysis are not limited to evolution education. The dynamic of introducing a fraudulent scientific controversy and legally codifying requirements to provide false information can be seen in recent abortion laws. Texas law, for example, requires doctors to provide a discredited scientific claim in order to suggest that abortion causes cancer. All women must listen to the following before they obtain an abortion: "Those opposed to abortion have attempted to link abortion with an increased risk of breast cancer. Numerous leading health organizations, including the American Cancer Society, have found through various studies that abortion does *not* increase the risk of breast cancer. *Please ask us about the natural protective effect of a completed pregnancy in mitigating the risk of breast cancer.*"[12] Robert Post, dean of Yale Law School, has argued that the Texas law violates doctors' first amendment rights because it requires them to recite false information to patients: no studies demonstrate a link between breast cancer and abortion.[13] Yet Alaska, Kansas, Mississippi, and Oklahoma also require written materials that assert a false link between breast cancer and abortion. If states can merely announce a controversy and then supply false information, they compromise the knowledge of citizens. A student, who has been taught in school to treat all theories as equal, will not be able to evaluate the claims in such a statement.

But there are also movements toward defending public reason in ways consistent with science for citizenship. Noam Scheiber, a senior editor at *The New Republic*, suggests that blogs like Nate Silver's 538 or his recent book *The Signal and the Noise: Why So Many Predictions Fail—but Some Don't* might support an evidence-based politics. Silver tries to understand why a candidate like Rick Perry can present the science of climate change—despite a vast consensus on certain key claims—as a hoax. Silver observes that one error on the part of the United Nations' International Panel on Climate Change (IPCC) has been used by climate skeptics to claim that all climate predictions are false. Silver's data demonstrates that the one error does not invalidate the IPCC's major claims. Scheiber views Silver as a public statistician "bringing simple but powerful empirical methods to bear on a controversial policy question, and making the results accessible to anyone with a high-school level of numeracy."[14] Silver does not present data as self-generating: a data set may be erroneous, and there is uncertainty involved in evaluating climate change or financial data. Even more important, Silver separates the data from interpretation: the job of asking the right policy questions. Although data helps clarify the issues and possible policy options, the data is not an end in itself: the issues are "relentlessly messy," and policy decisions require debate and deliberation.[15]

What is at issue in *Plyler* and eighteenth-century writing about education is the ability of the citizens, jurors, and voters to interpret the messy and the technical. Negotiating medical information or a debate about public policy requires something beyond basic literacy and numeracy: the ability to reason, ask questions, and evaluate. An informed discussion of global warming does not demand that we all be trained climate scientists, but liberal citizens need a basic understanding of science and how to evaluate evidence.

When liberals defensively exclude Intelligent Design without providing positive arguments for secular standards and science, they divert their energy from the real work: defining the explicit connections between science education and democracy. Fighting religion leaves liberalism at an impasse: a hollow victory in Dover satisfies the supporters of science, yet provides no defense against a religion-free ID that emphasizes academic freedom and "alternative views." While my analysis of ID only begins to describe the proper relationship between science and liberal politics, it embraces the Dover case as an opportunity to explore the relationship between education, science, and liberal citizenship. This work has focused on judicial review and political theory, but the language of democratic temperament and science for citizenship should be used in deliberation on legislation, interpretation of public opinion polls, and local debates about curriculum. The twentieth century began with Scopes and the exclusion of evolution because it conflicted with the narrative of Genesis. At the start of the twenty-first century, we see a shift to critical inquiry—so-called monkey bills—that demands a more explicit defense of American democracy as one that embraces public reason and trains citizens to deliberate. Liberals should choose to defend science education on these grounds. In the case of evolution education, there is a critical civic outcome that rests on making a choice—and defending it.

Notes

NOTES TO THE INTRODUCTION

1. Douglas O. Linder, *Famous Trials in History*, project of University of Missouri-Kansas City School of Law, http://law2.umkc.edu/faculty/projects/ftrials/ftrials.htm; Edward J. Larson, *Summer for the Gods: The Scopes Trial and America's Continuing Debate Over Science and Religion* (New York: Basic Books, 2006), chapters 6 and 7.
2. Amendment No. 1 to HB 368/SB 893, http://www.capitol.tn.gov/Bills/107/Amend/SA0901.pdf. Louisiana is the only other state that has successfully passed similar legislation. Similar bills have been introduced—and defeated—in Arkansas, Florida, Iowa, Kentucky, Maryland, Michigan, Mississippi, Missouri, New Hampshire, New Mexico, New York, Oklahoma, South Carolina, Texas, and Utah. Indiana is currently debating such a bill.
3. Nick Wing, "Tennessee Evolution Bill Becomes Law After Governor Bill Haslam Declines To Act," *The Huffington Post*, http://www.huffingtonpost.com/2012/04/10/tennessee-evolution-bill-haslam_n_1416015.html.
4. For examples of the ID texts, see William A. Dembski, *Intelligent Design: The Bridge Between Science & Theology* (Downers Grove, IL: InterVarsity Press, 2007); Michael J. Behe, *The Edge of Evolution: The Search for the Limits of Darwinism* (New York: Free Press, 2007); and Phillip E. Johnson, *The Wedge of Truth: Splitting the Foundations of Naturalism* (Downers Grove, IL: InterVarsity Press, 2000). The scientific community refutes ID's claims. See Kenneth R. Miller, *Only a Theory: Evolution and the Battle for America's Soul* (New York: Viking, 2008); Robert T. Pennock, ed., *Intelligent Design Creationism and Its Critics: Philosophical, Theological, and Scientific Perspectives* (Cambridge, MA: MIT Press, 2001); Laurie R. Godfrey, ed., *Scientists Confront Creationism* (New York: W.W. Norton & Company, 1983); National Research Council, *Science and Creationism: A View from the National Academy of Sciences, Second Edition* (Washington, DC: National Academy Press, 1999); Philip Kitcher, *Abusing Science: The Case Against Creationism* (Cambridge: MIT Press, 1982); Philip Kitcher, *Living with Darwin: Evolution, Design, and the Future of Faith* (Oxford: Oxford University Press, 2006); and Philip Kitcher, *Science, Truth, and Democracy* (Oxford: Oxford University Press, 2003).
5. J.P. Euben, "Political Science and Political Silence," in *Power and Community*, ed. Philip Green and Stanford Levinson (New York: Random House, 1971), 1–57; and Sheldon S. Wolin, " Political Theory as a Vocation," *The American Political Science Review* 63, no. 4 (Dec. 1969): 1062–82.

6. Lawrence S. Lerner et al., "The State of State Science Standards," Thomas Fordham Institute (January 31, 2012): 10, http://www.edexcellence.net/publications/the-state-of-state-science-standards-2012.html.
7. U.S. Department of Education, "The Nation's Report Card: Science 2009: National Assessment of Educational Progress at Grades 4, 8, and 12," *Institute of Education Sciences: National Center for Education Statistics*, January 2011, http://nces.edu.gov/nationsreportcard/pdf/main2009/2011451.pdf; PISA (Programme for International Student Assessment) http://www.oecd.org/unitedstates/presentationofthepisa2010results.htm.; TIMSS (Trends In International Mathematics and Science Study) http://timss.bc.edu/timss2007/intl_reports.html.
8. Lerner et al., "The State of State Science Standards," 9.
9. National Research Council, *Science, Evolution, and Creationism* (Washington, DC: The National Academy Press, 2008), 11.
10. Ibid.
11. Selman v. Cobb County School District, 449 F.3d 1320 (11th Cir. 2006), http://www.aibs.org/public-policy/evolution_state_news.html#15687. The court ruled that the stickers violated the Establishment Clause and must be removed. Prior to a retrial, the case was settled out of court with the school district agreeing that they would not attach stickers or other warnings.
12. For examples of the ID texts, see William A. Dembski, *Intelligent Design: The Bridge Between Science & Theology* (Downers Grove, IL: InterVarsity Press, 2007) and Michael J. Behe, *The Edge of Evolution: The Search for the Limits of Darwinism* (New York: Free Press, 2007).
13. There have been four extremely influential publications regarding the data to support evolution and the problems with creation science and Intelligent Design: National Academy of Sciences, *Teaching About Evolution and the Nature of Science* (Washington, DC: National Academy Press, 1998); National Research Council, *Science and Creationism: A View from the National Academy of Sciences, Second Edition* (Washington, DC: National Academy Press, 1999); National Academy of Sciences. *Evolution in Hawaii: A Supplement to Teaching About Evolution and the Nature of Science* (Washington, DC: National Academies Press, 2004); National Research Council, *Science, Evolution, and Creationism,* (Washington, DC: National Academy Press, 2008). In addition, several important works have reviewed ID claims and refuted them all: the scientific community rejects both the criticisms of evolutionary theory and the claim that Intelligent Design is science. See Miller, *Only a Theory*; Pennock, *Intelligent Design Creationism and Its Critics*; Godfrey, *Scientists Confront Creationism*; Kitcher, *Abusing Science*; Kitcher, *Living with Darwin*; and Kitcher, *Science, Truth, and Democracy*.
14. Amendment No. 1 to HB 368/SB 893, http://www.capitol.tn.gov/Bills/107/Amend/SA0901.pdf. See also the language of the bill's supporters: Sara Reardon, "Bill Allowing Teachers to Challenge Evolution Passes Tennessee House," *Science Insider*, April 7, 2011, http://news.sciencemag.org/scienceinsider/2011/04/bill-allowing-teachers-to-challenge.html; John Roach, "Intelligent design' in Tenn. schools?," *Cosmic Log on NBCNEWS.COM*, April 8, 2011, http://cosmiclog.msnbc.msn.com/_news/2011/04/08/6433919-intelligent-design-in-tenn-schools?lite; and Helen Thompson, "Tennessee 'Monkey Bill' Becomes Law," *Nature: International Weekly Journal of Science*, April 11, 2012, http://www.nature.com/news/tennessee-monkey-bill-becomes-law-1.10423.
15. Amendment No. 1 to HB 368/SB 893, http://www.capitol.tn.gov/Bills/107/Amend/SA0901.pdf.

16. See discussion in Chapter 1.
17. Based on the accounting of time in the Old and New Testaments, young earth creationists believe the earth is between 5,700 and 10,000 years old.
18. Frank Newport, "In U.S., 46% Hold Creationist View of Human Origins," Gallup Poll, June 1, 2012, http://www.gallup.com/poll/155003/Hold-Creationist-View-Human-Origins.aspx?version=print. For a more detailed analysis of American public opinion, see Chapter 2.
19. Alexis de Tocqueville, *Democracy in America*, trans George Lawrence ed. J.P. Mayer. (New York: Harper & Row, 1966).
20. John Rawls, *A Theory of Justice* revised edition (Boston: Harvard University Press, 2005).
21. Brighouse requires even more and links his claims to the autonomy of the student. Students must be able to identify "various sorts of fallacious arguments, and how to distinguish among them, as well as between them and non-fallacious arguments. The autonomous person needs to be able to distinguish between appeals to authority and appeals to evidence, between inductive and deductive arguments, as well as to identify ad hominem arguments and other misleading rhetorical devices." Harry Brighouse, *On Education* (London: Routledge, 2006), 24.
22. Susan Liebell, "Lockean Switching: Imagination and the Production of Principles of Toleration," *Perspectives on Politics* 7, no. 4 (December 2009): 823–36. Martha Nussbaum underlines the need for democratic citizens to have compassion for others—what she terms "narrative imagination"—in order to understand others and to be globally minded citizens: Martha C. Nussbaum, *Not for Profit: Why Democracy Needs the Humanities* (Princeton: Princeton University Press, 2010), 2, 6, 28–29, 39, 80, 95, and Chapter V. Brighouse requires citizens to engage in political participation in a spirit of respect and a willingness to engage in public reasoning (Brighouse, *On Education*, 67). See also John Rawls, *A Theory of Justice*, revised edition (Boston: Harvard University Press, 2005).
23. On how social contract theory may limit the participation of the mentally disabled, see Martha C. Nussbaum, *Frontiers of Justice: Disability, Nationality, Species Membership* (Cambridge: Harvard University Press, 2007).
24. Nussbaum and Brighouse reject economic *growth* as a democratic goal. Nussbaum agrees with the claims for the national interest and the individual (Nussbaum, *Not for Profit*, 9–10) as well as the creation of a business culture of innovation (53), while Brighouse emphasizes the role economic competence plays in individual flourishing (Brighouse, *On Education*, 28–41).
25. For example, Harry Brighouse, *On Education*; Eamonn Callan, *Creating Citizens: Political Education and Liberal Democracy* (Oxford: Oxford University Press, 2004); Meira Levinson, *The Demands of a Liberal Education* (Oxford: Oxford University Press, 2002); Stephen Macedo, *Diversity and Distrust: Civic Education in a Multicultural Democracy* (Cambridge: Harvard University Press, 2003); Martha C. Nussbaum, *Not for Profit*; Rob Reich, *Bridging Liberalism and Multiculturalism in American Education* (Chicago: University of Chicago Press, 2002).
26. Harry Brighouse, for example, believes that math and science are "great cultural achievements" that should be included in the curriculum because they assist individuals in making decisions related to leading a "good life" (Brighouse, *On Education*, 54).
27. Nussbaum, *Not for Profit*, 3, 5–6, 16–17, 20, 138. On endangerment, 7.
28. Amy Gutmann, *Democratic Education* (Princeton: Princeton University Press, 1999),102–3. See also Martha Nussbaum, *Not for Profit*.

NOTES TO CHAPTER 1

1. Massimo Pigliucci, *Nonsense on Stilts: How to Tell Science from Bunk* (Chicago: University of Chicago Press, 2010), 162–66.
2. Pigliucci, quoting Barbara Forest's trial testimony, *Nonsense on Stilts,* 166. See also Gordy Slack, *The Battle over the Meaning of Everything: Evolution, Intelligent Design, and a School Board in Dover, PA* (San Francisco: John Wiley & Sons, 2007) and Lauri Lebo, *The Devil in Dover: An Insider's Story of Dogma v. Darwin in Small-town America* (New York: New Press, 2008).
3. Edwards v. Aguillard, 482 U.S. 578 (1987).
4. Tammy Kitzmiller, et al. v. Dover Area School District, et al., 400 F. Supp. 2d 707 (M.D. Pa 2005), http://www.pamd.uscourts.gov/kitzmiller/kitzmiller_342.pdf, 2.
5. Pigliucci, *Nonsense on Stilts,* 172.
6. Pennsylvania State Code §235.10(2).
7. Pigliucci, *Nonsense on Stilts,* 172.
8. Tammy Kitzmiller, et al. v. Dover Area School District, et al., 400 F. Supp. 2d 707 (M.D. Pa 2005), http://www.pamd.uscourts.gov/kitzmiller/kitzmiller_342.pdf, 1–2.
9. James R. Moore, *The Post-Darwinian Controversies: A Study of the Protestant Struggle to Come to Terms with Darwin in Great Britain and America, 1870–1900* (Cambridge, England Cambridge University Press, 1979).
10. Michael Lienesch, *In the Beginning: Fundamentalism, The Scopes Trial, and the Making of the Anti-evolution Movement* (Chapel Hill: University of North Carolina Press, 2007), 8, 60, 69–73, 83.
11. Ibid., 21, 27–28, 92, 95–96, 221–24.
12. Ibid., 115n2, 176–77.
13. For accounts of the trial, see Edward J. Larson, *Summer for the Gods: The Scopes Trial and America's Continuing Debate Over Science and Religion* (New York: Basic Books, 2006) and Douglas O. Linder, "Stave v. Scopes: Trial Excerpts," *Famous Trials in History,* University of Missouri-Kansas City School of Law, http://law2.umkc.edu/faculty/projects/ftrials/scopes/scopes2.htm. See also H. L. Mencken, *A Religious Orgy in Tennessee: A Reporter's Account of the Scopes Monkey Trial* (New York: Melville House 2006).
14. John Scopes v. The State, 154 Tenn. 105, 289 S.W. 363 (Tenn. 1927).
15. Epperson v. Arkansas, 393 U.S. 97 (1968).
16. Daniels v. Waters, 515 F. 2d 485 (6th Cir. 1975).
17. McLean v. Arkansas Board of Education, 529 F.Supp. 1255 (ED Ark.1982).
18. Edwards v. Aguillard, 482 U.S. 578 (1987). Justices Powell and White wrote concurrences. Justice O'Connor did not join Part II.
19. *Edwards,*48 U.S. at 596 and 582 citing Don Aguillard, et al. v. Edwin Edwards, et.al., 765 F.2d 1251 (5th Cir. 1985); emphasis added. See also *Edwards,* 48 U.S. at 590–94, 596, 582–83.
20. *Edwards,* 48 U.S. at 586, 593.
21. Martha Nussbaum's discussion of ID misses the importance of the Scalia dissent. See Martha C. Nussbaum, *Liberty of Conscience: In Defense of America's Tradition of Religious Equality* (New York: Basic Books, 2008), 321, 324. As Thomas Nagel insists that science is too narrowly defined and should not make claims that religious claims are false, he mistakes criticism of evolution for support of ID and ends up supporting a position very much like Scalia's. He insists ID and creationism are different without sufficient proof and would include ID if it were a possibility. Thomas Nagel, "Public Education and Intelligent Design," *Philosophy & Public Affairs* 36, no. 2 (2008): 187–205.

22. *Edwards,* 48 U.S. at 628, 611, 634, 622.
23. Lienesch, *In the Beginning,* 30–31.
24. Larry Witham, "Many Scientists See God's Hand in Evolution," *Reports of the National Center for Science Education* 17, no. 6 (November–December 1997): 33, http://ncse.com/rncse/17/6/many-scientists-see-gods-hand-evolution; Henry Morris, "The ICR Scientists," *Impact* 86 (August 1980): i-viii.
25. This *amicus* was later published as *Science and Creationism: A View from the National Academy of Sciences,* and there have been several revisions (see Introduction, n12).
26. See Introduction, n13.
27. Jay B. Labov and Barbara Kline Pope, "Understanding Our Audiences: The Design and Evolution of *Science, Evolution, and Creationism,*" *The American Society for Cell Biology Education—Life Sciences Education* 7 (Spring 2008): 20–24, doi:10.1187/cbe.07–12–0103.
28. Justice Warren E. Burger, Lemon v. Kurtzman, 403 U.S. 602 (1971).
29. O'Connor, concurring in Lynch v. Donnelly, 465 U.S. 668, 688 (1984). Justices Scalia, Thomas, and Kennedy prefer "coercion" as the measure of government establishment of religion. See Kennedy, concurring and dissenting, County of Allegheny v. American Civil Liberties Union, 492 U.S. 573 (1989).
30. *Dover,* 400 F. Supp. 2d 19, quoting *McLean v. Arkansas,* 529 F. Supp. at 1258; *Dover* 400 F. Supp 2d 21; see also 19–21.
31. *Dover,* 400 F. Supp. 2d 19–35 (person), 36–50 (student), 50–64 (citizen); see also 49. Throughout his opinion, Jones refers to the ID movement as a unified group. ID is better understood as a coalition struggling with dissent over the strategic purging of references to God. Lienesch, *In the Beginning,* 236–37.
32. *Dover,* 400 F. Supp. 2d, 22–23; *Edwards,* 482 U.S. at 593.
33. Burger, *Lemon,* 403 US 602 (1971).
34. *Dover,* 400 F. Supp. 2d, 20–21, 32, 31–34.
35. Phillip Kitcher develops these terms in *Living with Darwin: Evolution, Design, and the Future of Faith* (Oxford: Oxford University Press, 2007), 7, 18, 81–107, 114.
36. *McLean,* 529 F. Supp. at 1267.
37. *Edwards,* 428 U.S. 578 (1987); Brief for 72 Nobel Laureates, 17 State Academies of Science, and 7 other Scientific Organizations as Amicus Curiae, Edwards v. Aguillard, 482 U.S 578 (1987).
38. *Dover,* 400 F. Supp. 2d 64, 65, 70, 75, 79, 82–84, 87–89. Clearly, Jones's concern with supernatural causation and research can be mapped onto *McLean's* concern with natural law and testability. Jones's third criterion—criticism of evolution has been refuted—attacks ID's negative thesis only.
39. *Dover,* 400 F. Supp. 2d 63; see also the fuller discussion of science at 63–89.
40. *Dover,* 400 F. Supp. 2d at 63–64; emphasis added.
41. *Edwards,* 482 U.S. 578 at 593.
42. Jones's historical summary can be found at *Dover,* 400 F. Supp. 2d at 18–19, 91–92.
43. *Dover,* 400 F. Supp. 2d at 42 citing *McLean,* 529 F. Supp. at 1266; see also 21, 49–50.
44. Jones does not use these more theoretical terms, but his discussion implies them. *Dover,* 400 F. Supp. 2d at 24–31, 71–75, 130–34.
45. The Tennessee law insists that it does not "promote any religious or non-religious doctrine, promote discrimination for or against a particular set of religious beliefs or non-beliefs, or promote discrimination for or against religion or non-religion." A court *might* claim that the disclaimer is the result of a fraught history of Christian literalists opposing evolution—and there is a

religious motive—but this would be a stretch given the literal meaning of the law. H.B.368/S.B. 893, http://www.capitol.tn.gov/Bills/107/Amend/SA0901. pdf.

46. *Dover*, 400 F. Supp. 2d at 93; see also 42, 68–69, 71.
47. Brief for 72 Nobel Laureates, 17 State Academies of Science, and 7 other Scientific Organizations as Amicus Curiae at 4, *Edwards v. Aguillard*, 482 U.S 578 (1987).

NOTES TO CHAPTER 2

1. William A. Galston, "Two Concepts of Liberalism," *Ethics* 105, no. 3 (April 1995): 516–34; William A. Galston, "Civic Education in the Liberal State," in *Liberalism and the Moral Life,* ed. Nancy Rosenblum (Cambridge, MA: Harvard University Press, 1989); Stephen Macedo, "Liberal Civic Education and Religious Fundamentalism: The Case of God v. John Rawls?," in *The Ethics of Citizenship: Liberal Democracy and Religious Convictions*, ed. J. Caleb Clinton (Waco, TX: Baylor University Press, 2009), 93–117; Brian Barry, *Culture and Equality: An Egalitarian Critique of Multicultural-ism* (Cambridge: Cambridge University Press, 2002); K. Anthony Appiah, "Liberal Education: The United States Example," in *Education and Citizenship in Liberal-Democratic Societies: Teaching for Cosmopolitan Values and Collective Identities,* ed. Kevin McDonough and Walter Feinberg (Oxford: Oxford University Press, 2003), 56–74, 69–71.
2. Susan Moller Okin, "Forty Acres and a Mule' for Women: Rawls and Feminism," *Politics, Philosophy & Economics* 4 (June 2005): 233–48; John Locke, "Second Treatise," in *Two Treatises of Government,* ed. Peter Laslett (Cambridge: Cambridge University Press, 1988); John Locke, *Some Thoughts Concerning Education ; and, Of the Conduct of the Understanding*, ed. Ruth Grant and Nathan Tarcov (Indianapolis, IN: Hackett Pub. Co., 1996); J.S. Mill, *On Liberty and Other Writings,* ed. Stefan Collini (Cambridge: Cambridge University Press, 1991); Barry, *Culture and Inequality*, 200.
3. Mill, *On Liberty*, 105, 4–108. Mill fears rather than favors the state educating all children, 106.
4. Barry, *Culture and Inequality*, 148, 203, 210; John Locke, "Second Treatise"; Mill, *On Liberty*; Ian MacMullen, *Faith in Schools? Autonomy, Citizenship, and Religious Education in the Liberal State* (Princeton: Princeton University Press, 2007), 3.
5. O'Connor, concurring in *Lynch v. Donnelly* 465 U.S. 668 (1984) at 688; see previous discussion, Chapter 1, 25–36.
6. Amy Gutmann, *Democratic Education* (Princeton: Princeton University Press, 1999), xiii.
7. Ibid., 95, 96, 98, 100.
8. Ibid., 101.
9. Ibid.
10. Ibid., 103.
11. Ibid., 102.
12. For example, Michael J. Behe, a professor of biological sciences at Lehigh University, presents arguments about cell complexity from biochemistry, while William A. Dembski, a professor in the conceptual foundations of science at Baylor University, provides the mathematical claims. Dembski has a Ph.D. in *both* mathematics and philosophy and a master of divinity in theology. William A. Dembski, *Intelligent Design: The Bridge Between Science & Theology* (Downers Grove, IL: InterVarsity Press, 2007) and Michael J.

Behe, *The Edge of Evolution: The Search for the Limits of Darwinism* (New York: Free Press, 2007).

13. Gutmann, *Democratic Education,* 102–3; Appiah, "Liberal Education: The United States Example," 71.
14. Eric Plutzer and Michael Berkman, "Evolution, Creationism, and the Teaching of Human Origins in Schools," *Public Opinion Quarterly* 72, no. 3 (Fall 2008): 540–53. Plutzer and Berkman analyze national polls from 1981–2007. The most recent Gallup Poll supports their conclusions: Frank Newport, "In U.S., 46% Hold Creationist View of Human Origins," *Gallup Poll,* June 1, 2012, http://www.gallup.com/poll/155003/Hold-Creationist-View-Human-Origins.aspx?version=print.
15. Plutzer and Berkman, "Evolution, Creationism, and the Teaching of Human Origins," 544, 548.
16. Ibid., 546–47.
17. "PRRI/RNS Religion News Survey September 14–18, 2011," *Public Religion Research Institute,* September 22, 2011, http://publicreligion.org/site/wp-content/uploads/2011/10/September-PRRI-RNS-Topline-Questionnaire-and-Survey-Methodology-.pdf. In the case of climate change, 84 percent of scientists accept humans caused climate change, while only 56 percent of the public agrees. The more educated support both evolution as well as climate change consensus at the highest rates. "Public Opinion on Religion and Science in the United States," *Pew Research Center for the People and the Press* (July 9, 2009), 38–39, http://www.people-press.org/files/legacy-pdf/528.pdf; and *Evolution and Creationism: A View from the National Academy of Sciences, 2nd Edition* (Washington, DC: National Academy Press, 1999), *FAQ,* 2–3; Jon D. Miller, Eugenie C. Scott, and Shinji Okamoto, "Science Communication: Public Acceptance of Evolution," *Science* 313 (2006): 765–66.
18. "Public Opinion on Religion and Science in the United States," 8, 20–21, 38–39. For further analysis of demographics in anti-evolution movements, see Simon Coleman and Leslie Carlin, eds., *The Cultures of Creationism: Anti-Evolution in English-Speaking Countries* (Aldershot, England: Ashgate, 2004).
19. "Public Opinion on Religion and Science in the United States," 37–39; Michael Berkman and Eric Plutzer, *Evolution, Creationism, and the Battle to Control America's Classrooms* (Cambridge: Cambridge University Press, 2010), 71–75.
20. Plutzer and Berkman, "Evolution, Creationism, and the Teaching of Human Origins in Schools," 550–51.
21. Miller, Scott, and Okamoto, "Science Communication: Public Acceptance of Evolution," 765 and Supporting Online material, 10. Only Turkey had lower acceptances of evolution than the United States.
22. Gutmann, *Democratic Education,* 103.
23. John Rawls, "The Idea of Public Reason Revisited," *The University of Chicago Law Review* 64, no. 3 (Summer 1997): 765–807.
24. Gutmann, *Democratic Education,* 102.
25. Barry, *Culture and Inequality,* 247; Stephen Holmes, *Passions and Constraints: On the Theory of Liberal Democracy* (Chicago: University of Chicago Press, 1987), 231.
26. Gutmann, *Democratic Education,* 102–4.
27. Harry Brighouse, *On Education* (London: Routledge, 2006), 2, 16–19; Appiah, "Liberal Education: The United States Example," 56–60, 71.
28. Amy Gutmann, "Civic Education and Social Diversity," *Ethics* 105, no. 3 (April 1995): 557–79, http://www.jstor.org.ezproxy.sju.edu/stable/2382142. See also Brighouse, *On Education,* 28–129.

29. Brighouse, *On Education,* 67; John Rawls, *A Theory of Justice,* Revised Edition (Cambridge, MA: University of Harvard Press, 1999), 99, 156, 397, 424, 436, 438.
30. Brighouse, *On Education,* 62.
31. Ibid., Chapter two, especially 28–41, 44.
32. Brighouse, *On Education,* 19, 24, 54. Children have a "right" to an education that prepares them for autonomy as adults, according to Appiah, "Liberal Education: The United States Example," 63.
33. Wisconsin v. Yoder, 406 U.S. 205 (1972), Justice William O. Douglas dissenting, at 242–47 and notes 3 (on the social science evidence demonstrating children's capacity to choose) and 4 (on the ability of the state to ask each student whether they want to go to high school); Appiah, "Liberal Education: The United States Example," 63. James G. Dwyer explores the child-centered approach to education in his work, for example, "Children's Right," in *A Companion to the Philosophy of Education,* ed. Randall Curren (Malden, MA: Blackwell, 2003), 450 and *Religious Schools v. Children's Rights* (Ithaca: Cornell University Press, 1998).
34. Barry, *Culture and Inequality,* 176–78, 192–93, 202–3; Jeff Spinner-Halev, *The Boundaries of Citizenship: Race, Ethnicity and Nationality in the Liberal State* (Baltimore: The Johns Hopkins University Press, 1994), 101–2, 110–11; Macedo calls this a real but constrained exit option, "Liberal Civic Education and Religious Fundamentalism," 111. Susan Moller Okin claims liberalism must ensure that the individual can exit while focusing on women's compromised exit rights that violate equality. Susan Moller Okin, "Mistresses of Their Own Destiny: Group Rights, Gender, and Realistic Rights of Exit," in *Education and Citizenship in Liberal-Democratic Societies: Teaching for Cosmopolitan Values and Collective Identities,* ed. Kevin McDonough and Walter Feinberg (Oxford: Oxford University Press, 2003), 325–50, 325–26.
35. Galston, "Two Concepts of Liberalism," 521–22, 533–34.
36. Ibid., 521.
37. Macedo, "Liberal Civic Education and Religious Fundamentalism," 94, 117.
38. Rawls, *A Theory of Justice,* 86, 190–94.
39. Macedo, "Liberal Civic Education and Religious Fundamentalism," 94, 95, 101, 105, 108, 112, 117.
40. Many religious conservatives believe that schools are a source of danger and experience a sense of persecution when evolution or other subjects are taught to their children. See Berkman and Plutzer, *Evolution, Creationism, and the Battle to Control America's Classrooms,* 60–61.
41. Mozert v. Hawkins County Board of Education, 827 F.2d 1058 (6th Cir. 1987).
42. Macedo, "Liberal Civic Education and Religious Fundamentalism," 97. For an excellent critique of political liberalism's obsession with *Mozert,* see Alisa Kessel, "The New Parochialism?: Liberal education and the danger of teaching tolerance," 2012, unpublished manuscript.
43. Macedo, "Liberal Civic Education and Religious Fundamentalism," 97, 98, 99, 100, 107, 108, 109, 112, 418n59.
44. Ibid., 100. On how art affects critical thinking, see Susanne K. Langer, *Philosophy in a New Key: A Study in the Symbolism of Reason, Rite, and Art* (Cambridge: Harvard University Press, 1996) and Maxine Greene, "Prologue to Art, Social Imagination and Action," *Journal of Educational Controversy* 5, no. 1 (Winter 2010), 1, http://www.wce.wwu.edu/Resources/CEP/eJournal/v005n001/p001.shtml

45. Barry, *Culture and Inequality*, 248; Appiah, "Liberal Education: The United States Example," 56–74, 59–61.
46. Macedo, "Liberal Civic Education and Religious Fundamentalism," 95n59, 418. For his critique of science, Macedo relies upon Stanley Fish, "Liberalism Doesn't Exist," *Duke Law Journal* 36, no. 6 (1987): 997–1001.
47. Macedo, "Liberal Civic Education and Religious Fundamentalism," 103–5, 114n48, 417. For more nuanced interpretations of Dewey, see Melvin I. Rogers, *The Undiscovered Dewey: Religion, Morality, and the Methods of Democracy* (New York: Columbia University Press, 2009) and Jack Knight and James Johnson, *The Priority of Democracy: Political Consequences of Pragmatism* (New York: Russell Sage Foundation, 2011).
48. Macedo, "Liberal Civic Education and Religious Fundamentalism," 103–4.

NOTES TO CHAPTER 3

1. Organizations may submit briefs to the Court that suggest arguments outside of those made by the plaintiff and defendant.
2. Philip B. Kurland and Ralph Lerner, *The Founders' Constitution* (Chicago: University of Chicago Press, 1987), volume 3 citing Max Farrand, ed. *The Records of the Federal Convention of 1787*. Rev. ed. 4 vols. New Haven and London: Yale University Press, 1937, 2:321, Journal, 18 Aug; 2:505, Journal, 5 Sept.; 2:595; Committee of Style.
3. James Madison, "Journal of the Constitutional Convention of 1787," in *The Writings of James Madison, Comprising His Public Papers and His Private Correspondence, Including his Numerous Letters and Documents Now for the First Time Printed*, ed. Gaillard Hunt (New York: G.P. Putnam's Sons, 1900), vol. 3, http://oll.libertyfund.org/title/1935/118621/2395726.
4. Thomas Jefferson, *Notes on the State of Virginia*, 1782, Q.XIV, ME 2:206–7; Thomas Jefferson to James Madison, 1787, FE 4:480; *Diffusion of Knowledge Bill*, 1779, FE 2:221, Papers 2:527; ME 7:253. Later in his career, Jefferson further championed education as crucial to the stability and long-term success of popular government. See discussion in Chapter 5.
5. Rush Welter, *Popular Education and Democratic Thought in America* (New York: Columbia University Press, 1962), 24–25. The language of these constitutions is discussed in more detail in the next chapter.
6. Welter, *Popular Education*, 10–22; William J. Reese, *America's Public Schools: From the Common School to "No Child Left Behind"* (Baltimore: The Johns Hopkins University Press, 2005), 1–44; Carl F. Kaestle, *Pillars of the Republican: Common Schools and American Society, 1780–1860* (New York: Hill and Wang, 1983), 1–12. See also Bernard Bailyn, *Education in the Forming of American Society: Needs and Opportunities for Study* (Chapel Hill: University of North Carolina Press, 1960); Lawrence A. Cremin, *American Education: The Colonial Experience 1607–1783* (New York: Harper 1970).
7. Reese, *America's Public Schools*, 23–62.
8. Ibid., 68–72.
9. Ibid., 77.
10. Ibid., 5.
11. Thomas D. Snyder and Alexandra G. Tan, *Digest of Education Statistics, 2004* (Washington, D.C.: U.S. Government Printing Office, 2005); Michael S. Katz, *A History of Compulsory Education Laws. Fastback Series, No. 75. Bicentennial Series* (Bloomington, IN: Phi Delta Kappa, 1976).

12. Reese, *America's Public Schools*, 3–4.
13. Wisconsin v. Yoder, 406 U.S. 205 (1972) and Ambach v. Norwick, 441 U.S. 68 (1979). See also Meyer v. Nebraska 262 U.S. 390 (1923); Brown v. Board of Education 347 U.S. 483 (1954); Abington School District v. Schempp 374 U.S. 203 (1963).
14. *Yoder,* 406 U.S. at 221.
15. *Yoder,* 406 U.S. at 222.
16. Plyler v. Doe 457 U.S. 202 (1982).
17. *Plyler,* 457 U.S. at 221, citing Abington School District v. Schempp, 374 U.S. at 230 (Brennan concurring) and Meyer v. Nebraska 262 U.S. 390, 400 (1923).
18. *Plyler,* 457 U.S. at 223 citing *Brown* 347 U.S. at 493.
19. *Plyler,* 457 U.S. at 221 citing *Yoder,* 406 U.S. at 221.
20. *Plyler,* 457 U.S. at 221 citing *Ambach,* 441 U.S. at 77–78: "These perceptions of the public schools as inculcating fundamental values necessary to the maintenance of a democratic political system have been confirmed by the observations of social scientists. See R. Dawson & K. Prewitt, Political Socialization 146–167 (1969); R. Hess & J. Torney, The Development of Political Attitudes in Children 114, 158–171, 217–220 (1967); V. Key, Public Opinion and American Democracy 323–343 (1961)."
21. *Plyler,* 457 U.S. at 222.
22. *Plyler,* 457 U.S. at 221, citing San Antonio Independent School Dist. v. Rodriguez, 411 U.S. 1, 35 (1973).
23. *Plyler,* 457 U.S. at 221, 222, 224.
24. *Plyler,* 457 U.S. at 222, emphasis added.
25. *Plyler,* 457 U.S. at 221.
26. "no state shall . . . deny to any person within its jurisdiction the equal protection of the laws."
27. *Yoder,* 406 U.S. at 211. See discussion in chapter 2, 35–36.
28. *Plyler,* 457 U.S. at 221; emphasis added.
29. *Plyler,* 457 U.S. at 223, citing *Brown,* 347 U.S. at 493.
30. *Plyler,* 457 U.S. at 222.
31. Council on Foreign Relations, *U.S. Education Reform and National Security,* 7, http://www.cfr.org/united-states/us-education-reform-national-security/p27618.
32. Ibid., 54. Science as essential for modern industry (ix, x, xiii, 3, 4); inability of employers to find skilled employees (42); Americans less scientifically capable compared to other nation's graduates (ix, 10, 17, 23); lack of science and technology as endangering national security (x, 46) and the ability to find skilled military officers (4, 9); decline in patents (4).
33. Ibid., 17.
34. Harry Brighouse, *On Education* (London: Routledge, 2006), 28 and chapter 2. See my discussion above, chapter 2.
35. Ibid., 62.
36. Ibid., 4.
37. This approach remains today. States set standards that are locally administered. Even No Child Left Behind—a national mandate—forces assessment and improvement rates rather than content or methods. Increasingly, test administration creates some grey. Although there are no teaching methods proscribed by NCLB, the common practice of tailoring teaching methods to tested materials prioritizes tested content above anything else: an indirect mandate to teach specific content.
38. *Plyler* 231 citing *San Antonio Independent School District v. Rodriguez,* 411 U.S. 1, (1973), 111.

39. Marshall quoting William v. Rhodes 393 U.S. 23, 32 (1968). John Stuart Mill, *On Liberty and Other Writings*, ed. Stefan Collini (London: Cambridge University Press, 1989), 31, 47. Oliver Wendall Holmes uses the marketplace of ideas in his famous dissent in Abrams v. U.S. 250 U.S. 616, 630 (1919). See also Brennan in Lamont v. Postmaster General 381 U.S. 308, 301 (1965). John Milton's *Areopagitica* (1644) also depends on the marketplace of ideas as a liberal solution for discussion and making government smarter.
40. *San Antonio Independent School District*, 411 U.S. 1, Marshall dissenting, 71, 102, 110–14, 117. Marshall quotes from Williams v. Rhodes, 393 U.S. 23, 32 (1968) and Schenck v. United States, 249 U.S. 47 (1919). Marshall's concurrence in *Plyler* is heavily based on this earlier opinion.
41. The right to vote, never explicitly stated, may be regulated by the states but may not be abridged based on "race, color, or previous condition of servitude" (15th Amendment), sex (19th Amendment), failure to pay poll or other taxes (24th Amendment), or age (all 18 year olds must vote, 26th amendment).
42. *Plyler*, 457 U.S. at 233, citing Reynolds v. Sims, 377 U.S. 533, 562 (1964).
43. *San Antonio Independent School District*, 411 U.S. at 114, citing *Reynolds*, 377 U.S. 533.While the studies Marshall cites are dated, two recent articles support Marshall. Appealing to civic education theory, Hillygus claims a direct correlation between voting rates and taking the SAT/attending a four-year college, while Burden argues that education influences turnout and the effect of college education on voting increases in the 1980s. Burden, however, maintains that education levels—not merely education—affect voting with the more highly educated voting more. D. Sunshine Hillygus, "The Missing Link: Exploring the Relationship Between Higher Education and Political Education and Political Engagement," *Political Behavior* 27, no. 1 (March 2005): 25–47; Barry C. Burden, "The Dynamic Effects of Education on Voter Turnout," *Electoral Studies* 28, no. 4 (December 2009): 540–49.
44. Brian Barry, *Culture and Equality: An Egalitarian Critique of Multiculturalism* (Cambridge: Cambridge University Press, 2002), 212–13.
45. *San Antonio Independent School District*, 411 U.S. 1, 115.
46. *San Antonio Independent School District*, 411 U.S. 1, 112 citing Sweezy v. New Hampshire, 354 U.S. 234 (1957), 250.
47. *San Antonio Independent School District*, 411 U.S. at 112–13; emphasis added.
48. *Yoder*, 406 U.S. at 245.
49. The public and privates goods distinction is handled well by Rob Reich and William S. Koski, "The State's Obligation to Provide Education: Adequate Education or Equal Education?" (unpublished manuscript, American Political Science Association, 2007).
50. John Adams presents a very similar understanding of rights in *Dissertation on the Feudal and Canon Law*, and his ideas are considered in Chapter 5.
51. Campaign for Fiscal Equity v. State, 100 N.Y.2d 893 at 951 (2003), 801 N.E.2d 326.
52. For example, National Research Council, "Evidence Supporting Biological Evolution," in *Science and Creationism: A View from the National Academy of Sciences, 2nd ed.* (Washington D.C.: The National Academy Press, 1999), 5–9, http://www.nap.edu/catalog.php?record_id=6024.
53. Brief for 72 Nobel Laureates, 17 State Academies of Science, and 7 other Scientific Organizations as Amicus Curiae, *Edwards v. Aguillard*, 482 U.S. 578 (1987).
54. Ibid.

55. Brief for National Academy of Sciences as Amicus Curiae, Edwards v. Aguillard, 482 U.S 578 (1987, http://www.soc.umn.edu/~samaha/cases/edwards_v_aguillard_NAC.html.
56. Ibid.
57. *McLean v. Arkansas Board of Education,* 529 F.Supp. 1255 (ED Ark.1982) at 1267; Tammy Kitzmiller, et al. v. Dover Area School District, et al., 400 F. Supp. 2d 707 (M.D. Pa 2005), http://www.pamd.uscourts.gov/kitzmiller/kitzmiller_342.pdf, 63,
58. Brief for 72 Nobel Laureates, 17 State Academies of Science, and 7 other Scientific Organizations as Amicus Curiae, *Edwards v. Aguillard,* 482 U.S 578 (1987).
59. Stephen Breyer, *Active Liberty: Interpreting Our Democratic Constitution* (New York: Alfred A. Knopf, 2005).
60. Ibid., 8–9.
61. Breyer does not need an appeal to ancient liberty, Benjamin Constant, or Pocockian historiography to assert that the constitutional requires citizens, at the very least, able to present tyranny. Breyer mentions Adams only a few times (3, 21–22, 135), but Adams desires education as a public good (tyranny prevention, protection of free press) and a private good (enlightenment of the individual). John Adams's position is discussed in Chapter 5.
62. Breyer, *Active Liberty,* 46–47, 57–58, 64–66, 70–71, 101.

NOTES TO CHAPTER 4

1. John Dewey, *Democracy and Education: An Introduction to the Philosophy of Education* (New York: The Free Press, 1916), 219.
2. Ibid., 219–21; Max Fisch, ed., *Classic American Philosophers* (New York: Fordham University Press, 1995); in particular, John Dewey, *Science and Society,* 387; *The Construction of Good,* 373; and *The Supremacy of Method,* 348–49; 357.
3. Stephen Gaukroger argues for an earlier emergence of science. Stephen Gaukroger, *The Emergence of a Scientific Culture: Science and the Shaping of Modernity, 1210–1685* (Oxford: Clarendon Press, 2006).
4. Fisch (Dewey), *Supremacy of Method,* 350–57; Dewey, *Democracy and Education,* 225; *The Influence of Darwinism,* 340.
5. Dewey recounts a similar movement in the early twentieth century; *The Influence of Darwinism,* 340–41.
6. *The Influence of Darwinism,* 343; *The Construction of Good,* 367–71, 373, 378; *Science and Society,* 382–83, 388; *Democracy and Education,* 223–27.
7. *Democracy and Education,* 224, 228–30. The National Academy of Sciences continues to emphasize the contribution of science to "technological innovation," "cures for disease," and knowledge necessary to produce "labor-saving devices." National Research Council, *Science, Evolution, and Creationism* (Washington, D.C.: The National Academy Press, 2008), preface, p. 1, http://www.nap.edu/html/creationism/non-javascript/preface.html
8. *Democracy and Education,* 224, 228; *Science and Society,* 381–83, 387; *The Construction of Good,* 372; *Science and Society,* 389; *Creative Democracy,* 392–93.
9. Harry Brighouse, *On Education* (London: Routledge, 2006), 24, 212.
10. Susan Liebell, "Lockean Switching: Imagination and the Production of Principles of Toleration," *Perspectives on Politics* 7, No. 4 (December 2009):

823–36. See also Martha Nussbaum's concept of "narrative imagination" for democratic citizenship, Martha C. Nussbaum, *Not for Profit: Why Democracy Needs the Humanities* (Princeton: Princeton University Press, 2010), 2, 6, 28–29, 39, 80, 95, and chapter V; as well as Brighouse's requirement of respect and a willingness to engage in public reasoning, Brighouse, *On Education*, 67, 122, based on John Rawls, *A Theory of Justice*, Revised Edition (Cambridge, MA: University of Harvard Press, 1999). Barry favors education to present lucid arguments as well as to help students to understand "hygiene and public health in order to practice effective contraception and to raise children properly," but the United States lacks consensus on what raising children "properly" requires. Brian Barry, *Culture and Equality: An Egalitarian Critique of Multiculturalism* (Cambridge: Cambridge University Press, 2001), 212, 224, 228. For a discussion of the issues raised by sex education, see Josh Corngold, "Toleration, Parents' Rights, and Children's Autonomy: The Case of Sex Education," UMI Dissertation Publishing (September 3, 2011); and Nomi Maya Stolzenberg, " 'He Drew a Circle that Shut Me Out': Assimilation, Indoctrination, and the Paradox of a Liberal Education," *Harvard Law Review*, 106, no 3 (January1993), 581-667.

11. On how social contract theory may limit the participation of the mentally disabled, see Martha C. Nussbaum, *Frontiers of Justice: Disability, Nationality, Species Membership* (Cambridge: Harvard University Press, 2007).
12. Brian Barry, *Culture and Equality*, 212.
13. Nussbaum and Brighouse reject economic *growth* as a democratic goal. Nussbaum agrees with the claims for the national interest and the individual, as well as the creation of a business culture of innovation: Nussbaum, *Not for Profit*, 9–10, 53; while Brighouse emphasizes the role economic competence plays in individual flourishing: Brighouse, *On Education*, chapter 2, especially 28–41.
14. Barry, *Culture and Equality*, 212.
15. Ian MacMullen, *Faith in Schools? Autonomy, Citizenship, and Religious Education in the Liberal State* (Princeton: Princeton University Press, 2007), 3.
16. Martha Craven Nussbaum, *Cultivating Humanity: A Classical Defense of Reform in Liberal Education* (Cambridge, MA: Harvard University Press, 1997), 290. On the relationship between biology and education, see Martha C. Nussbaum, *Not for Profit*, 74, 82, 181–82, 187, 226, 243–44, 253. Her conflation of technology and science is best seen at 3, 5–6, 16–17, 20, 138.
17. Nussbaum, *Cultivating Humanity*, 8–12.
18. K. Anthony Appiah, "Liberal Education: The United States Example," in *Education and Citizenship in Liberal-Democratic Societies: Teaching for Cosmopolitan Values and Collective Identities*, ed. Kevin McDonough and Walter Feinberg (Oxford: Oxford University Press, 2003), 56–74.
19. Brian Barry includes science in the education that would encourage this discourse: Barry, *Culture and Equality*, 222–25, 229, 238.

NOTES TO CHAPTER 5

1. John Adams, "Diary: With Passages from an Autobiography," in Vol. 2 of *The Works of John Adams, Second President of the United States: With a Life of the Author, Notes and Illustrations, by* Charles Francis Adams (Boston: Little, Brown and Co., 1856), http://oll.libertyfund.org/title/2100/159580.

2. Scott M. Liell, "Shaking the Foundation of Faith," *The New York Times*, November 18, 2005; Lawrence A. Cremin, *American Education: The Colonial Experience 1607–1783* (New York: Harper, 1970), 514–15.

3. John Winthrop, the fourth generation of Winthrops in Massachusetts, taught science and math at Harvard for 40 years. He was with a friend of Washington, Franklin, and Adams. Lawrence A. Cremin, *American Education: The Colonial Experience 1607–1783* (New York: Harper, 1970), 514–15.

4. Cremin, *American Education,* 416.

5. John Adams, "Dissertation on the Feudal Canon Law," in *A Collection of State-Papers, Relative to the First Acknowledgment of the Sovereignty of the United States of America* (1765), Project Gutenberg, January 2010, ¶15, http://www.gutenberg.org/fi les/30872/30872-h/30872-h.htm. For all references, I will use the paragraph rather than the page number.

6. Adams, "Dissertation," ¶1.

7. Adams, "Dissertation," ¶2. While Kant also sees social benefit, he emphasizes the individual experience of knowing and changing. See Immanuel Kant, *What is Enlightenment?* (1784).

8. Adams, "Dissertation," ¶2. Compare to John Locke, *Second Treatise,* §225: "long train of abuses," which would "rouze" people to rebellion against a monarch.

9. Adams uses language common in the eighteenth century for the working poor and working class.

10. When Brennan and Marshall address education for immigrant groups, they are far less comfortable addressing issues of class. Instead, they focus on the exclusion of minority groups.

11. Adams, "Dissertation," ¶3, emphasis added.

12. Adams, "Dissertation," ¶3.

13. Adams, "Dissertation," ¶14 and 15; Aristotle, *The Politics*, 1282b.

14. Adams, "Dissertation," ¶28, 15.

15. Adams, "Dissertation," ¶14.

16. Adams, "Dissertation," ¶15.

17. Adams, "Dissertation," ¶28.

18. Adams, "Dissertation," ¶15.

19. Adams, "Dissertation," ¶10, 14. For a brief overview of the exclusion of Catholics in American liberal thinking, see Rogers M. Smith, *Civic Ideals* (New Haven: Yale University Press, 1997), 54–58 and 72–74.

20. Bernard Bailyn, *Education in the Forming of American Society: Needs and Opportunities for Study* (Chapel Hill: University of California Press, 1960), 26–27; Cremin, *American Education,* 124–25. See previous discussion in Chapter 3.

21. Alexis de Tocqueville, *Democracy in America*, trans George Lawrence, ed. J.P. Mayer (New York: Harper & Row, 1966).

22. Rogers Smith notes that the rough equality among that group was juxtaposed with assumed inequalities between men and women, northern Europeans and African or Native Americans, gays and heterosexuals, and Protestants and Catholics, Jews, or Muslims. Smith, *Civic Ideals*, 17.

23. See Bernard Bailyn, *Education in the Forming of American Society*, 27; Cremin, *American Education*; and Carl F. Kaestle, *Pillars of the Republican: Common Schools and American Society, 1780–1860* (New York: Hill and Wang, 1983).

24. For example, Adams seems to have rejected the notion that all people who *lived* in the colonies and were subject to British rule were British subjects: "When James Otis instead asserted, consistent with Coke, that all 'colonists,

black and white' born here, are free born British subjects,' John Adams was one of many who "shuddered." See Smith, *Civic Ideals*, 73.

25. Bailyn, *Education in the Forming of American Society*, 41, 11–21, 45.
26. Adams, "Dissertation," ¶24.
27. Adams, "Dissertation," ¶24.
28. Adams, "Dissertation," ¶28. George Washington uses similar language in his Farewell Address, calling for the "the diffusion of knowledge." Bernard I. Cohen, *Science and the Founding Fathers: Science in the Political Thought of Jefferson, Franklin, Adams and Madison* (New York: W.W. Norton, 1995), 13.
29. Adams, "Dissertation," ¶25–27.
30. Martin Clagett, *Scientific Jefferson Revealed* (Charlottesville: University of Virginia Press, 2009), 59–63; Keith Thomson, *A Passion for Nature: Thomas Jefferson and Natural History* (Monticello: Thomas Jefferson Foundation, 2008), 79, 122; and Silvio A. Bedini, *Jefferson and Science* (Monticello: Thomas Jefferson Foundation, 2002), 57–64, 92–99. See also James Gilreath, ed., *Thomas Jefferson and the Education of a Citizen* (Washington, D.C.: Library of Congress, 1999) and James Conant, *Thomas Jefferson and the Development of American Public Education* (Berkeley: University of California Press 1963). General resources on Jefferson and education (including unpublished holdings from Monticello) are catalogued at the Thomas Jefferson Portal: http://tjportal.monticello. org/cgibin/Pwebrecon.cgi?DB=local&SL=none&SAB1=jefferson+educa tion+democra%3F&BOOL1=all+of+these&FLD1=Title%2C+Author+ %26+Subject+%28TASS%29&GRP1=AND+with+next+set&SAB2=& BOOL2=all+of+these&FLD2=Keyword+Anywhere+%28GKEY%29& CNT=50.
31. Thomas Jefferson to William Duane, 1810, *The Writings of Thomas Jefferson: Memorial Edition*, ed. Albert Ellery Bergh and Andrew Adgate Lipscomb (Washington D.C.: Issued under the auspices of the Thomas Jefferson Memorial Association of the United States, 1903–1904), 12:417 (hereafter ME followed by volume and page); Thomas Jefferson, "A Bill for the More General Diffusion of Knowledge, 1779," in *The Writings of Thomas Jefferson*, ed. Paul Leicester Ford (G. P Putnam's Sons, New York: 1892–1899) 2:221 (hereafter FE followed by volume and page); Thomas Jefferson: Notes on Virginia Q.XIV, 1782, ME 2:207; Thomas Jefferson to William C. Jarvis, 1820, ME 15:278; Thomas Jefferson to Richard Price, 1789, ME 7:253; Thomas Jefferson: 1st Inaugural Address, 1801, ME 3:322; Thomas Jefferson: Opinion on Apportionment Bill, 1792, ME 3:211.
32. Thomas Jefferson to Peter Carr, 1814, ME 19:213; Thomas Jefferson: Reply to American Philosophical Society, 1808.
33. Thomas Jefferson to Littleton Waller Tazewell, 1805; Thomas Jefferson to Joseph C. Cabell, 1814, ME 14:84; Thomas Jefferson to James Madison, 1787, ME 6:392; Thomas Jefferson to James Madison, 1787, FE 4:480.
34. Thomas Jefferson to George Washington, 1786, ME 19:24; Thomas Jefferson to Joseph C. Cabell, 1818, FE 10:102; Thomas Jefferson: A Bill for the More General Diffusion of Knowledge, 1779, FE 2:221, Papers 2:527; Thomas Jefferson to James Madison, 1787, ME 6:392; Thomas Jefferson to James Madison, 1787, FE 4:480.
35. Thomas Jefferson copying Montesquieu's *Spirit of the Laws* IV, Ch. 5 into his Commonplace Book.
36. Thomas Jefferson to Pierre Samuel Dupont de Nemours, 1816, ME 14:491; Thomas Jefferson to Chevalier de Ouis, 1814, ME 14:130.

37. Thomas Jefferson: Report of the Commissioners for the University of Virginia, 1818, Electronic Text Center, University of Virginia Library, http://etext.virginia.edu/toc/modeng/public/JefRock.html.
38. Thomas Jefferson: A Bill for the More General Diffusion of Knowledge, 1779, FE 2:221; Thomas Jefferson to Hugh L. White, et al., 1810, ME 12:388.
39. Thomas Jefferson: Notes on Virginia Q.XIV, 1782, ME 2:206 and Thomas Jefferson to Mann Page, 1795, ME 9:30. Thomas Jefferson: Autobiography, 1821, ME 1:54.
40. Thomas Jefferson to Joseph C. Cabell, 1818, FE 10:99; Thomas Jefferson to Amos J. Cook, 1816, ME 14:405; Thomas Jefferson to Samuel Knox, 1810, ME 12:360; Thomas Jefferson to James Breckinridge, 1821, ME 15:314; Thomas Jefferson to Cornelius Camden Blatchly, 1822, ME 15:399; Thomas Jefferson to M. A. Jullien, 1818, ME 15:172.
41. Thomas Jefferson: A Bill for the More General Diffusion of Knowledge, 1779, FE 2:221.
42. Thomas Jefferson to Benjamin Waterhouse, 1818, FE, 12:89–90.
43. Thomas Jefferson to Roger C. Weightman, 1826, ME 16:182.
44. Thomas Jefferson to Peter Carr, August 19, 1785, in *The Papers of Thomas Jefferson*, ed. Julian P. Boyd et al. (Princeton: Princeton University Press, 1950), 8:405–6.
45. Thomas Jefferson to——, 1821, ME 15:340.
46. Thomas Jefferson to John Adams, 1814, in *Thomas Jefferson Writings*, ed. Merrill Peterson (New York: Library of America, 1984), 1343.
47. Thomas Jefferson, "*Memorial on the Book Duty*: November 30, 1821," in *Public Papers* of *Thomas Jefferson 1743–1826*, University of Virginia Library Electronic Text Center, http://etext.virginia.edu/toc/modeng/public/JefPapr.html.
48. Ibid.
49. Ibid.
50. Ibid.
51. Thomas Paine, "The Age of Reason," in *The Writings of Thomas Paine*, vol. 4, ed. Moncure Daniel Conway (New York: G.P. Putnam's Sons, 1894), Kindle edition, Locations 331, 587, 615, 1013, 1025, 1079, 2677.
52. Ibid., Locations 305, 697, 722, 2726, 2675.
53. Ibid., Locations 626, 687, 704–15.
54. Ibid., Location 2681.
55. Ibid., Location 743.
56. Ibid., Locations 2722, 2714. 843.
57. Ibid., Locations 626, 664, 685, 722, 741, 743, 791, 843, 2681, 2686, 2694, 2714, 2722.
58. Ibid., Location 630. See also Paine's rejection of the learning of dead languages in favor of scientific knowledge at 774.
59. Ibid., Locations 2701, 2582.
60. Ibid., Location 2710.
61. Ibid., Locations 1013, 1025, 2025, 2601, 2659.
62. Ibid., Location 305.
63. Ibid., Location 791.
64. Ibid., Locations 418, 435, 454, 524, 553, 560, 561, 674, 690, 1025, 1039, 2601, 2649, 2659, 2694, 2705, 2707.
65. I. Bernard Cohen, *Science and the Founding Fathers*, 196 and 198–210.
66. Ibid., 26–27 and 225–30.
67. Adams, "Diary."
68. Adams, "Dissertation," ¶28.
69. Ibid., 23.

NOTES TO CHAPTER 6

1. Paul Broun, speaking at Liberty Baptist Church, Hartwell Georgia, "U.S. Rep. Paul Broun: Evolution a lie 'from the pit of hell,'" *Los Angeles Times*, September 27, 2012, http://articles.latimes.com/2012/oct/07/nation/la-na-nn-paul-broun-evolution-hell-20121007; Hillary Mayell, "Oldest Human Fossils Identified," *National Geographic News*, February 16, 2005, http://news.nationalgeographic.com/news/2005/02/0216_050216_omo.html
2. Richard Dawkins, *The Selfish Gene* (Oxford: Oxford University Press, 1976); *The Blind Watchmaker* (Burnt Mill, Harlow, Essex, England: Longman Scientific and Technical, 1986); *River out of Eden: A Darwinian View of Life* (London: Phoenix, 1995); *The God Delusion* (Boston: Houghton Mifflin, 2006). See also Daniel Dennett, *Darwin's Dangerous Idea: Evolution and the Meaning of Life* (New York: Simon and Shuster, 1995) and *Breaking the Spell: Religion as a Natural Phenomenon* (New York: Viking, 2006). Peter Bowler observes that Dawkins and Dennett are appealing to the tradition of Huxley, discussed later: Peter Bowler, *Monkey Trials and Gorilla Sermons* (Cambridge: Harvard University Press, 2007), 4, 76–77, 103–10.
3. John Hedley Brooke and Ronald L. Numbers, eds., *Science and Religion Around the World* (Oxford: Oxford University Press, 2011), 1–3, 92–119.
4. Thomas Dixon, *Science and Religion: A Very Short Introduction* (Oxford: Oxford University Press, 2008), Kindle edition, Location 509, 517. For more on the relationship between science and religion, see Ian G. Barbour, *Science and Religion: New Perspectives on the Dialogue* (New York: Harper and Row, 1968); John Hedley Brooke, *Science and Religion: Some Historical Perspectives* (Cambridge: Cambridge University Press, 1991), ch. 6; John R. Durant, *Darwinism and Divinity: Essays on Evolution and Religious Belief* (Oxford: Basil Blackwell, 1985); Gary B. Ferngren, ed. *Science and Religion: A Historical Introduction* (Baltimore: Johns Hopkins University Press, 2002); John C. Greene, *Debating Darwin: Adventures of a Scholar* (Claremont, CA: Regina Books, 1999); David C. Lindbert and Ronald I. Numbers, eds., *God and Nature: Historical Essays on the Encounter between Science and Religion* (Berkeley: University of California Press, 1986); Ronald L. Numbers, *The Creationists: From Scientific Creationism to Intelligent Design*, expanded edition (Cambridge, MA: Harvard University Press, 2006), ch. 17; Paul Wood, ed., *Science and Dissent in England, 1688–1945* (Aldershot, UK: Ashgate, 2004); Peter J. Bowler, *Reconciling Science and Religion* (Chicago: University of Chicago Press, 2001); Ian G. Barbour, *Religion in an Age of Science* (London: SCM Press, 1990), chs. 5–7; John F. Haught, *God after Darwin: A Theology of Evolution* (Boulder, CO: Westview Press, 2000); and Arthur Peacocke, *Creation of the World of Science* (Oxford: Oxford University Press, 2004).
5. Geoffrey Cantor, "Modern Judaism," in *Science and Religion Around the World*, ed. John Hedley Brooke and Ronald L. Numbers (Oxford: Oxford University Press, 2011), 44–66; Noah Efron, "Early Judaism," in *Science and Religion Around the World*, ed. John Hedley Brooke and Ronald L. Numbers (Oxford: Oxford University Press, 2011), 20–43. See also *Guide to the Perplexed*, Book II, Chapter 25 and "Letter on Astrology," in *A Maimonides Reader*, ed. Isadore Twersy (New York: Behrman House, 1972), 463–73; Norbert Samuleson, "Judaism and Science," in *The Oxford Handbook of Religion and Science*, ed. Philip Clayton (Oxford: Oxford University Press, 2006), 41–44.
6. Cantor, "Modern Judaism," 44–66, 0–51.
7. For example, Gerald L. Schroeder, *The Science of God: The Convergence of Scientific and Biblical Wisdom* (New York: Broadway Books, 1998) and

Nathan Aviezer, *In the Beginning: Biblical Creation and Science* (Jersey City: Ktav, 1990).

8. Natan Slifkin, "A General Response to the Charge of Heresy," http://www.zootorah.com/controversy/scienceresponse.html; Natan Slifkin, *The Challenge of Creation: Judaism's Encounter with Science, Cosmology and Evolution* (Springfield, NJ: Geffen Books, 2012); Natan Slifkin, *The Science of Torah: The Reflections of Torah in the laws of Science, the Creation of the Universe, and the Development of Life* (Southfield, MI: Targum, 2001). See also Geoffrey Cantor and Marc Swetlitz, eds., *Jewish Tradition and the Challenge of Darwinism* (Chicago: University of Chicago Press, 2006), 208–24.

9. For detail on exegesis of Qur'anic references to nature, see Ahmad S. Dullal, "Early Islam," in *Science and Religion Around the World*, ed. John Hedley Brooke and Ronald L. Numbers (Oxford: Oxford University Press, 2011), 138–42.

10. For example, al-Kindi (801–873), al-Farabi (870–950), Ibn-Khaldun (1332–1406). Ibid., 120–47.

11. For example, al-Biruni (973–1048), Ibn al-Haytham (965–1040), Ibn al-Shatir (1304–1375) (astronomy); al-Khayyam (1048–1131), Sharaf al-Din al-Tusi (1135–1213) (mathematics); Ibn al-Haytham, (965–1040) (optics); Avicenna or Ibn Sina (980–1037) (medicine); Ibid, 126–33, 145. On astronomy and Islam, see David King, *Astronomy in the Service of Islam* (Aldershot, UK: Variorum, 1993) and *Islamic Mathematical Astronomy* (London: Variorum, 1986).

12. S. J. Badakhchani, "Nasir al-Din Tusi," *Internet Encyclopedia of Philosophy*, http://www.iep.utm.edu/tusi/. The *Book of Animals* was published in seven volumes in Cairo (1323–1324) and reprinted in 1887. See also Mehmet Bayrakdar, "Al-Jahiz And the Rise of Biological Evolutionism," *The Islamic Quarterly*, vol. 27, no. 3 (1983), 149-155. Al-Dinawari (828–896) and Ibn Khaldun (1332–1406) are also often cited as examples. See also Dixon, *Science and Religion*, Locations 1239, 1240, 1321, 1329.

13. Muẓaffar Iqbāl, *Science and Islam* (New York: Greenwood, 2007), 158. Ismail Mazhar (1891–1962) translated the chapters, but a complete translation did not appear until 1964.

14. Ekmeleddin Ihsanoğlu, "Modern Islam," in *Science and Religion Around the World*, ed. John Hedley Brooke and Ronald L. Numbers (Oxford: Oxford University Press, 2011), 164–65.

15. Ibid., 148–74, 163.

16. Draper's *Conflict between Religion and Science* (1874) was published in French (1875) and Turkish (1895, 1897, 1900).

17. Ihsanoğlu, "Modern Islam," 165.

18. Najm A. Bezirgan, "The Islamic World," in *The Comparative Reception of Darwin*, ed. Thomas Glick (Chicago: University of Chicago Press, 1988), 375–87.

19. Albert Hourani, *Arabic Thought in the Liberal Age, 1789–1939* (London: Oxford University Press, 1962), 248–53.

20. Ihsanoğlu, "Modern Islam," 166; Hourani, *Arabic Thought in the Liberal Age*, 222–24.

21. Hârun Yahya, *The Evolution Deceit: The Scientific Collapse of Darwinism and Its Ideological Background* (London: Ta-Ha, 1999); Ihsanoğlu, "Modern Islam," 170–71.

22. Ibid., 168–70. Rooted in the work of Bediuzzarman Said Nursi (1877–1960), this approach remains popular. See, for example, Maurice Bucaille's *The Bible, the Qur'an, and Science* (Indianapolis: American Trust Publications, 1978). On Said Nursi, see Ibrahim Kahn, "Three Views of Science in

the Islamic World," in *God, Life, and the Cosmos: Christian and Islamic Perspectives*, ed. Ted Peters, Muzaffar Iqbal, and Syed Normanul Haq (Aldershot, UK: Ashgate, 2002), 43–75. On science and Islam, see Ziyauddin Sardar, *Islamic Futures: The Shape of Ideas to Come* (London: Mansell, 1985) and Ziyauddin Sardar, *Explorations in Islamic Science* (London: Mansell, 1989).

23. See the discussion of Fundamentalism in Chapter 1.
24. John Heilbron, *The Sun in the Church: Cathedrals as Solar Observatories* (Cambridge, MA: Harvard University Press, 1999). On the Jesuits, see Mordechai Feingold, ed., *The New Science and Jesuit Science: Seventeenth-Century Perspectives* (Dordrecht: Kluwer, 2003); Mordechai Feingold, ed., *Jesuit Science and the Republic of Letters* (Cambridge, MA: MIT Press, 2002); Marcus Hellyer, *Catholic Physics: Jesuit Natural Philosophy in Early Modern Germany* (Notre Dame, IN: University of Notre Dame Press, 2005); Peter Harrison and David C. Lindberg, "Early Christianity," in *Science and Religion Around the World*, ed. John Hedley Brooke and Ronald L. Numbers (Oxford: Oxford University Press, 2011), 82.
25. Augustine, *Literal Commentary on Genesis*, trans. John Hammon Taylor, in *Ancient Christian Writers*, vol. 41 (New York: Newman Press 1982), 42–43. See also David C. Lindberg, "The Medieval Church Encounters the Classical Tradition: Saint Augustine, Roger Bacon, and The Handmaiden Metaphor," in *When Science and Christianity Meet*, ed. David C. Lindberg and Ronald L. Numbers (Chicago: University of Chicago Press, 2008), 7–32.
26. John Hedley Brooke, "Modern Christianity," in *Science and Religion Around the World*, ed. John Hedley Brooke and Ronald L. Numbers (Oxford: Oxford University Press, 2011), 93. Brooke's work on Christianity is almost exclusively directed at Protestantism. For more on Catholicism, see Don O'Leary, *Roman Catholicism and Modern Science: A History* (New York: Continuum, 2007) and William B. Ashworth, Jr., "Catholicism and Early Modern Science," in *God and Nature: Historical Essays on the Encounter between Christianity and Science*, ed. David C. Lindberg and Ronald L. Numbers (Berkeley: University of California Press, 1986), 136–66, 147.
27. Rene Descartes, Letter to Mersenne [April 15, 1630], trans. Marjorie Grene and Roger Ariew, in *Rene Descartes: Philosophical Essays and Correspondence*, ed. Roger Ariew (Indianapolis: Hackett, 2000), 28.
28. Dixon, *Science and Religion*, Locations 827–30; Harrison and Lindberg, "Early Christianity," 79–80. Newton believed that God must intervene to keep planetary order: God as a watchmaker who occasionally winds or mends the watch. G. W. Leibniz preferred a God who acted once, with perfect foresight, to create a perfect and complete world. Dixon, *Science and Religion*, Location 811.
29. Dixon, *Science and Religion*, Location 1228.
30. John Henry Newman, Letter to J. Walker of Scarborough, May 22, 1868, *The Letters and Diaries of John Henry Newman* (Oxford: Clarendon Press, 1973).
31. For example, Père M. D. Leroy's *L'évolution restreinte aux espèces organiques* (1887) and John Augustine Zahm's *Evolution and Dogma* (1896) were placed on the Index in 1895 and 1898, respectively. See O'Leary, *Roman Catholicism and Modern Science*, 94–100; R. Scott Appleby, *Between Americanism and Modernism: John Zahm and Theistic Evolution*, in *Critical Issues in American Religious History: A Reader*, 2nd revised edition, ed. Robert R. Mathisen (Waco: Baylor University Press, 2006).
32. Dixon, *Science and Religion*, Location 1228.

33. Benedict XVI, "Mass Imposition of the Pallium and Conferral of the Fisherman's Ring for the Beginning of the Petrine Ministry of the Mass," Homily given at St. Peter's Square, April 24, 2005, http://www.vatican.va/holy_father/benedict_xvi/homilies/2005/documents/hf_ben-xvi_hom_20050424_inizio-pontificato_en.html.

34. Scottish evangelical Thomas Chalmers (1780–1847), Oxford geologist William Buckland (1784–1856), and American Congregational minister Edward Hitchcock (1793–1864) all favored a gap between "in the beginning" and the first "day" in Genesis to account for the differences. Geologist Hugh Miller (1802–1856) argued that the fossil record roughly followed the sequence in Genesis with geological epochs as days. Committed Christian scientists who did not see a conflict between a better understanding of laws of nature (particularly in thermodynamics) and belief in God include Michael Faraday and William Thomson; Brooke, "Modern Christianity," 104–6.

35. For example, a volume of the publication *Essays and Reviews* (1860) contained essays by several Anglican clergy who argued that the Bible should be read like any other book.

36. Brooke, "Modern Christianity," 105–6.

37. Dixon, *A Short History*, Location 237. See also Peter Bowler, *Monkey Trials and Gorilla Sermons: Evolution and Christianity from Darwin to Intelligent Design* (Cambridge: Harvard University Press, 2009).

38. Tess Cosslett, *Science and Religion in the Nineteenth Century* (Cambridge (Cambridge University Press, 1984), 152.

39. Charles Darwin, "Letter to Asa Gray," May 22, 1860, *Darwin Correspondence Database*,?http://www.darwinproject.ac.uk/entry-2814.

40. J.R. Lucas, "Wilberforce and Huxley: A Legendary Encounter," *The Historical Journal* 22, no. 2 (June 1979): 313–30; Dixon, *A Short History*, Locations 1148, 1161–64, 1187, 1193–98.

41. Dixon, *Science and Religion*, Chapter 3 and Locations 790.

42. Ibid., Chapter 3 and Locations 790, 798, 796, 802, 1145; and Brooke, "Modern Christianity," 108–9. See Asa Gray, *Darwiniana*, ed. A. Huner Dupree (Cambridge, MA: Harvard University Press, 1963), 72–145 and 293–320; and also David N. Livingstone, *Darwin's Forgotten Defenders* (Grand Rapids, MI: Eerdmans, 1987), 60–64. Charles Kingsley (1819–1875), Anglican Priest and writer, also supported Darwin. Hedley Brooke, "Modern Christianity," 108.

43. Dixon, *A Short History*, Locations 970–74.

44. "Resolution 2006-A129: Affirm Evolution and Science Education," *The Archives of the Episcopal Church, Acts of Convention 1976–2009*, http://www.episcopalarchives.org/cgi-bin/acts/acts_resolution-complete.pl?resolution=2006-A129.

45. Ibid. See also William G. Pollard, *Chance And Providence; God's Action In A World Governed By Scientific* (New York: Scribner, 1958); Arthur Peacocke, *Evolution: The Disguised Friend of Faith?* (West Conshohocken, PA: Templeton Foundation Press, 2004); John Polkinghorne, *Exploring Reality: The Intertwining of Science and Religion* (New Haven: Yale University Press, 2007).

46. "Resolution 1982-D090," *The Archives of the Episcopal Church, Acts of Convention 1976–2009*, http://www.episcopalarchives.org/cgi-bin/acts/acts_resolution.pl?resolution=1982-D090.

47. "Resolution 2006-A129: Affirm Evolution and Science Education," emphasis added. See also John Hedley Brooke and Geoffrey Cantor, *Reconstructing Nature: The Engagement of Science and Religion* (Edinburgh: T and T Clark, 1998), chs. 5–7; and Peter Harrison, *The Bible, Protestantism, and*

the Rise of Natural Science (Cambridge: Cambridge University Press, 1998), chs. 5–6.

48. John Hedley Brooke, "Modern Christianity," in *Science and Religion Around the World*, 110.

49. Richard Conniff, "God and White Men at Yale," *Yale Alumni Magazine* LXXV, No. 5 (May/June 2012): 44–51. See also Daniel Kevles, *In the Name of Eugenics: Genetics and the Uses of Human Heredity* (Cambridge: Harvard University Press, 1998); James R. Moore, "Ronald Aylmer Fisher: A Faith Fit for Eugenics," in *Eminent Lives in Twentieth-Century Science & Religion,* ed. Nicholas A. Rupke (New York: Peter Lang, 2007), 103–38, 118.

50. "From Francis Galton to George W. Hunter: Breaking Dogmatic Barriers and the Rise of the Eugenics Movement" by Douglas O. Linder (2005), http://law2.umkc.edu/faculty/projects/ftrials/conlaw/eugenics.html. Bryan used a particularly problematic quotation from *The Descent of Man* in which Darwin compared natural selection ("weak in body or mind are soon eliminated") to modern asylums, poor laws, and hospitals. He writes, "There is reason to believe that vaccination has preserved thousands who, from a weak constitution, would have succumbed to smallpox. Thus the weak members of civilized society propagate their kind. No one who has attended to the breeding of domestic animals will doubt that this must be highly injurious to the race of man. It is surprising how soon . . . care wrongly directed leads to the degeneration of a domestic race." Charles Darwin, *The Descent of Man, and Selection in Relation to Sex* (New York: D. Appleton, 1872), 162. On Bryan and the trial, see Edward J. Larson, *Summer of the Gods: The Scopes Trial and America's Continuing Debate Over Science and Religion* (New York: Basic Books, 2006), 28, 115, 245; as well as Edward J. Larson, "The Scopes Trial in History and Legend," in *When Science and Christianity Meet,* ed. David C. Lindberg and Ronald L. Numbers (Chicago: University of Chicago Press, 2008), ch. 11.

51. Dixon, *Science and Religion*, Locations 2, 92–119.

52. See discussion in Chapter 1.

53. See Laurie Goodstein, "New Light for Creationism," *New York Times*, December 21, 1997, sec. 4, 1, 4.

54. Richard Dawkins, "Obscurantism to the Rescue," *Quarterly Review of Biology*, 72 (1997), 397.

55. Michael Ruse agrees that polarization does not forward science education. Michael Ruse, *The Darwinian Revolution: Science Red in Tooth and Claw* (Chicago: University of Chicago Press, 1979); *Monad to Man: The Concept of Progress in Evolutionary Biology* (Cambridge, MA: Harvard University Press, 1996); *Can a Darwinian Be a Christian? The Relationship between Science and Religion* (Cambridge, MA: Harvard University Press, 2001); *Darwin and Design: Does Evolution Have a Purpose?* (Cambridge, MA: Harvard University Press, 2003); *The Evolution-Creation Struggle* (Cambridge: Harvard University Press, 2005).

NOTES TO CHAPTER 7

1. John F. Kennedy, "Science as a Guide of Public Policy," Address to the National Academy of Sciences, Washington, D.C. (October 22, 1963); Walter McDougall, *The Heavens and the Earth: A Political History of the Space Age* (New York: Basic Books, 1985).

2. Corey Dade, "In Their Own Words: GOP Candidates And Science," NPR, September 7, 2011, http://www.npr.org/2011/09/07/140071973/in-their-own-words-gop-candidates-and-science.

3. Michael Berkman and Eric Plutzer, *Evolution, Creationism, and the Battle to Control America's Classrooms* (Cambridge: Cambridge University Press, 2010), 35, 37.

4. Richard Dawkins, Book Review of Donald Johanson and Maitland Edey's *Blueprint, The New York Times*, section 7, April 9, 1989; Richard Dawkins, *The God Delusion* (Boston: Houghton Mifflin, 2006). Also see Chapter 6, note 2.

5. Ray Ginger, *Six Days or Forever: Tennessee v. John Thomas Scopes* (Oxford: Oxford University Press 1958), 129.

6. Dawkins, Book Review of Donald Johanson and Maitland Edey's *Blueprint*; Richard Dawkins, *The God Delusion* (Boston: Houghton Mifflin, 2006).

7. "Evolution, Creation, Intelligent Design," Gallup Poll, January 30, 2013, http://www.gallup.com/poll/21814/evolution-creationism-intelligent-design.aspx

8. Larry Witham, "Many Scientists See God's Hand in Evolution," *Reports of the National Center for Science Education*, vol. 17, issue 6, 1997 (Nov–Dec), 33 at http://ncse.com/rncse/17/6/many-scientists-see-gods-hand-evolution

9. Stephen Gaukroger, *The Emergence of a Scientific Culture: Science and the Shaping of Modernity: 1210–1685* (Oxford: Clarendon Press, 2006), 1–41, 145–52.

10. Ruth W. Grant, "Political Theory, Political Science, and Politics," *Political Theory* 30, no. 44 (August 2002): 584–85.

11. Jeremy Elkins, "Revolutionary Politics," *Theory & Event* 9, no. 4 (2006) and "Concerning Practices of Truth," in *Truth and Democracy*, ed. Jeremy Elkins and Andrew Norrish (Philadelphia: University of Pennsylvania Press, 2012), 19–53, 36–47.

12. *Texas Right to Know Act*, 2003. Italics supplied. Health and Safety Code, Sec. 171.012, http://www.statutes.legis.state.tx.us/Docs/HS/htm/HS.171.htm.

13. Robert C. Post, "Informed Consent to Abortion: A First Amendment Analysis of Compelled Physician Speech" (2007), Faculty Scholarship Series, Paper 170, http://digitalcommons.law.yale.edu/fss_papers/170; Elaine Schattner, "Do Abortions Cause Breast Cancer? The Shaky Science Behind Kansas' House Abortion Act," *Slate* (May 23, 2012), http://www.slate.com/articles/health_and_science/medical_examiner/2012/05/do_abortions_cause_breast_cancer_kansas_state_house_abortion_act_invokes_shaky_science_for_political_gain_.html.

14. Noam Scheiber, "Known Unknowns," *The New York Times Book Review*, Sunday November 4, 2012, p. 12;

15. Ibid.; Nate Silver, *The Signal and the Noise: Why So Many Predictions Fail— but Some Don't* (New York: Penguin Press, 2012).

Bibliography

Adams, John. "Diary: With Passages from an Autobiography." In *The Works of John Adams, Second President of the United States: With a Life of the Author, Notes and Illustrations*, edited by Charles Francis Adams. Boston: Little, Brown and Co., 1856. Accessed May 26, 2011. http://oll.libertyfund.org/title/2100/159580.

Adams, John. "Dissertation on the Feudal Canon Law." In *A Collection of State-Papers, Relative to the First Acknowledgment of the Sovereignty of the United States of America*, 2010. Project Gutenberg. Accessed August 18, 2009. http://www.gutenberg.org/files/30872/30872-h/30872-h.htm.

Appiah, K. Anthony. "Liberal Education: The United States Example." In *Education and Citizenship in Liberal-Democratic Societies: Teaching for Cosmopolitan Values and Collective Identities*, edited by Kevin McDonough and Walter Feinberg, 56–74. Oxford: Oxford University Press, 2003.

Appleby, Scott. "Between Americanism and Modernism; John Zahm and Theistic Evolution." In *Critical Issues in American Religious History: A Reader*, edited by Robert R. Mathisen, 2nd revised edition. Waco: Baylor University Press, 2006.

Ashworth, William B. Jr. "Catholicism and Early Modern Science." In *God and Nature: Historical Essays on the Encounter Between Christianity and Science*, edited by David C. Lindberg and Ronald L. Numbers. Berkeley: University of California Press, 1986.

Augustine. "*Literal Commentary on Genesis.*" In *Ancient Christian Writers*, vol. 41, translated by John Hammon Taylor. New York: Newman Press, 1982.

Aviezer, Nathan. *In the Beginning: Biblical Creation and Science*. Jersey City: Ktav, 1990.

Badakhchani, S.J. "Nasir al-Din Tusi." *Internet Encyclopedia of Philosophy*. Accessed August 1, 2012. http://www.iep.utm.edu/tusi/.

Bailyn, Bernard. *Education in the Forming of American Society: Needs and Opportunities for Study*. Chapel Hill: University of North Carolina Press, 1960.

Barbour, Ian G. *Religion in an Age of Science*. London: SCM Press, 1990.

Barbour, Ian G. *Science and Religion: New Perspectives on the Dialogue*. New York: Harper and Row, 1968.

Barry, Brian. *Culture and Equality: An Egalitarian Critique of Multiculturalism*. Cambridge: Cambridge University Press, 2002.

Bayrakdar, Mehmet. "Al-Jahiz And the Rise of Biological Evolutionism." *The Islamic Quarterly*, vol. 27, no. 3 (1983), 149–155.

Bedini, Silvio A. *Jefferson and Science*. Monticello: Thomas Jefferson Foundation, 2002.

Behe, Michael J. *The Edge of Evolution: The Search for the Limits of Darwinism*. New York: Free Press, 2007.

Benedict XVI. "Mass Imposition of the Pallium and Conferral of the Fisherman's Ring for the Beginning of the Petrine Ministry of the Mass," April 24, 2005. Accessed August 7, 2012. http://www.vatican.va/holy_father/benedict_xvi/homilies/2005/documents/hf_ben-xvi_hom_20050424_inizio-pontificato_en.html.

Bergh, Albert Ellery, and Andrew Adgate Lipscomb, eds. *The Writings of Thomas Jefferson: Memorial Edition* (ME). Washington, D.C.: Issued under the auspices of the Thomas Jefferson Memorial Association of the United States, 1903–1904.

Berkman, Michael, and Eric Plutzer. *Evolution, Creationism, and the Battle to Control America's Classrooms*. Cambridge: Cambridge University Press, 2010.

Bezirgan, Najm A. "The Islamic World." In *The Comparative Reception of Darwin*, edited by Thomas Glick, 375–87. Chicago: University of Chicago Press, 1988.

Bowler, Peter J. *Monkey Trials and Gorilla Sermons*. Cambridge, MA: Harvard University Press, 2007.

Bowler, Peter J. *Reconciling Science and Religion*. Chicago: University of Chicago Press, 2001.

Boyd, Julian P., Charles T. Cullen, John Catanzariti, Barbara B. Oberg, et al., eds. *The Papers of Thomas Jefferson*. Princeton: Princeton University Press, 1950.

Breyer, Stephen. *Active Liberty: Interpreting Our Democratic Constitution*. New York: Alfred A. Knopf, 2005.

Brighouse, Harry. *On Education*. London and New York: Routledge, 2006.

Brooke, John Hedley. "Modern Christianity." In *Science and Religion Around the World*, edited by John Hedley Brooke and Ronald L. Numbers, 92–119. Oxford: Oxford University Press, 2011.

Brooke, John Hedley. *Science and Religion: Some Historical Perspectives*. Cambridge: Cambridge University Press, 1991.

Brooke, John Hedley, and Geoffrey Cantor. *Reconstructing Nature: The Engagement of Science and Religion*. Edinburgh: T and T Clark, 1998.

Brooke, John Hedley, and Ronald L. Numbers, eds. *Science and Religion Around the World*. Oxford: Oxford University Press, 2011.

Bucaille, Maurice. *The Bible, the Qur'an, and Science*. Indianapolis: American Trust Publications, 1978.

Burden, Barry C. "The Dynamic Effects of Education on Voter Turnout." *Electoral Studies* 28, no. 4 (December 2009): 540–49.

Callan, Eamonn. *Creating Citizens: Political Education and Liberal Democracy*. Oxford: Oxford University Press, 2004.

Cantor, Geoffrey, and Marc Swetlitz, eds. *Jewish Tradition and the Challenge of Darwinism*. Chicago: University of Chicago Press, 2006.

Clagett, Martin. *Scientific Jefferson Revealed*. Charlottesville: University of Virginia Press, 2009.

Cohen, I. Bernard. *Science and the Founding Fathers: Science in the Political Thought of Jefferson, Franklin, Adams and Madison*. New York: W.W. Norton, 1995.

Coleman, Simon, and Leslie Carlin, eds. *The Cultures of Creationism: Anti-Evolution in English-Speaking Countries*. Aldershot, UK: Ashgate, 2004.

Conant, James. *Thomas Jefferson and the Development of American Public Education*. Berkeley: University of California Press, 1963.

Conniff, Richard. "God and White Men at Yale." *Yale Alumni Magazine* LXXV, no. 5 (May/June 2012): 44–51.

Corngold, Josh. *Toleration, Parents' Rights, and Children's Autonomy: The Case of Sex Education*. ProQuest, UMI Dissertation Publishing (September 3, 2011).

Council on Foreign Relations. *U.S. Education Reform and National Security*. Accessed January 27, 2013. http://www.cfr.org/united-states/us-education-reform-national-security/p27618.

Cremin, Lawrence A. *American Education: The Colonial Experience 1607–1783*. New York: Harper, 1970.

Dade, Corey. "In Their Own Words: GOP Candidates And Science." NPR, September 7, 2011. Accessed January 2, 2013. http://www.npr.org/2011/09/07/140071973/in-their-own-words-gop-candidates-and-science.

Darwin, Charles. *The Descent of Man, and Selection in Relation to Sex.* New York: D. Appleton, 1872.

Darwin, Charles. "Letter to Asa Gray, May 22, 1860." *Darwin Correspondence Database.* Accessed on January 31, 2013. http://www.darwinproject.ac.uk/entry-2814.

Dawkins, Richard. *The Blind Watchmaker.* Harlow: Longman Scientific and Technical, 1986.

Dawkins, Richard. *The God Delusion.* Boston: Houghton Mifflin, 2006.

Dawkins, Richard. "Obscurantism to the Rescue." *Quarterly Review of Biology,* 72 (1997): 397.

Dawkins, Richard. Review of Donald Johanson and Maitland Edey's *Blueprint. The New York Times.* April 9, 1989.

Dawkins, Richard. *River out of Eden: A Darwinian View of Life.* London: Phoenix, 1995.

Dawkins, Richard. *The Selfish Gene.* Oxford University Press, 1976.

Demski, William A. *Intelligent Design: The Bridge Between Science & Theology.* Downers Grove, IL: InterVarsity Press, 2007.

Dennett, Daniel. *Breaking the Spell: Religion as a Natural Phenomenon.* New York: Viking, 2006.

Dennett, Daniel. *Darwin's Dangerous Idea: Evolution and the Meaning of Life.* New York: Simon and Shuster, 1995.

Descartes, Rene. "Letter to Mersenne April 15, 1630." In *Rene Descartes: Philosophical Essays and Correspondence,* edited by Roger Ariew and translated by Marjorie Grene and Roger Ariew. Indianapolis: Hackett, 2000.

de Tocqueville, Alexis. *Democracy in America,* translated by George Lawrence and edited by J.P. Mayer. New York: Harper & Row, 1966.

Dewey, John. *Democracy and Education: An Introduction to the Philosophy of Education.* New York: The Free Press, 1916.

Dewey, John. *Science and Society.* In *Classic American Philosophers,* edited by Max Fisch. New York: Fordham University Press, 1995.

Dixon, Thomas. *Science and Religion: A Very Short Introduction.* Oxford: Oxford University Press, 2008, Kindle edition.

Dullal, Ahmad S. "Early Islam." In *Science and Religion Around the World,* edited by John Hedley Brooke and Ronald L. Numbers, 120–47. Oxford: Oxford University Press, 2011.

Durant, John R. *Darwinism and Divinity: Essays on Evolution and Religious Belief.* Oxford: Basil Blackwell, 1985.

Dwyer, James G. "Children's Right." In *A Companion to the Philosophy of Education,* edited by Randall Curren. Malden, MA: Blackwell, 2003.

Dwyer, James G. *Religious Schools v. Children's Rights.* Ithaca: Cornell University Press, 1998.

Elkins, Jeremy. "Concerning Practices of Truth." In *Truth and Democracy,* edited by Jeremy Elkins and Andrew Norrish, 19–53. Philadelphia: University of Pennsylvania Press, 2012.

Elkins, Jeremy. "Revolutionary Politics." *Theory & Event* 9, no. 4 (2006).

Euben, J.P. "Political Science and Political Silence." In *Power and Community,* edited by Philip Green and Stanford Levinson, 1–57. New York: Random House, 1971.

Feingold, Mordechai, ed. *Jesuit Science and the Republic of Letters.* Cambridge: MA: MIT Press, 2002.

Feingold, Mordechai, ed. *The New Science and Jesuit Science: Seventeenth-Century Perspectives.* Dordrecht: Kluwer, 2003.

Ferngren, Gary B., ed. *Science and Religion: A Historical Introduction*. Baltimore: Johns Hopkins University Press, 2002.

Fish, Stanley. "Liberalism Doesn't Exist," *Duke Law Journal* (1987): 997–1001.

Ford, Paul Leicester, ed. *The Writings of Thomas Jefferson* (FE). G. P Putnam's Sons, New York: 1892–1899.

Galston, William A. "Civic Education in the Liberal State." In *Liberalism and the Moral Life*, edited by Nancy Rosenblum. Cambridge, MA: Harvard University Press, 1989.

Galston, William A. "Two Concepts of Liberalism." *Ethics* 105, no. 3 (Apr., 1995): 516–34.

Gaukroger, Stephen. *The Emergence of a Scientific Culture: Science and the Shaping of Modernity, 1210–1685*. Oxford, Clarendon Press, 2006.

Gilreath, James, ed. *Thomas Jefferson and the Education of a Citizen*. Washington, D.C.: Library of Congress, 1999.

Ginger, Ray. *Six Days or Forever: Tennessee v. John Thomas Scopes*. Oxford: Oxford University Press, 1958.

Godfrey, Laurie R., ed. *Scientists Confront Creationism*. New York and London: W.W. Norton & Company, 1983.

Goodstein, Laurie. "New Light for Creationism." *New York Times*, December 21, 1997.

Grant, Ruth W. "Political Theory, Political Science, and Politics." *Political Theory* 30, no. 4 (August 2002): 584–85.

Gray, Asa. *Darwiniana*, edited by A Huner Dupree. Cambridge, MA: Harvard University Press, 1963.

Greene, John C. *Debating Darwin: Adventures of a Scholar*. Claremont, CA: Regina Books, 1999.

Greene, Maxine. "Prologue to Art, Social Imagination and Action," *Journal of Educational Controversy* 5, no. 1 (Winter 2010), 1. http://www.wce.wwu.edu/Resources/CEP/eJournal/v005n001/p001.shtml

Gutmann, Amy. "Civic Education and Social Diversity." *Ethics* 105, no. 3 (Apr., 1995): 557–79. Accessed January 7, 2010. http://www.jstor.org.ezproxy.sju.edu/stable/2382142.

Gutmann, Amy. *Democratic Education*. Princeton: Princeton University Press, 1999.

Harrison, Peter and David C. Lindberg. "Early Christianity." In *Science and Religion Around the World*, edited by John Hedley Brooke and Ronald L. Numbers, 67–92. Oxford: Oxford University Press, 2011.

Harrison, Peter. *The Bible, Protestantism, and the Rise of Natural Science*. Cambridge: Cambridge University Press, 1998.

Haught, John F. *God after Darwin: A Theology of Evolution*. Boulder, CO: Westview Press, 2000.

Heilbron, John. *The Sun in the Church: Cathedrals as Solar Observatories*. Cambridge, MA: Harvard University Press, 1999.

Hellyer, Marcus. *Catholic Physics: Jesuit Natural Philosophy in Early Modern Germany*. Notre Dame, IN: University of Notre Dame Press, 2005.

Hillygus, D. Sunshine. "The Missing Link: Exploring the Relationship Between Higher Education and Political Education and Political Engagement." *Political Behavior* 27, no. 1 (March 2005): 25–47.

Holmes, Stephen. *Passions and Constraints: On the Theory of Liberal Democracy*. Chicago: University of Chicago Press, 1987.

Hourani, Albert. *Arabic Thought in the Liberal Age, 1789–1939*. London: Oxford University Press, 1962.

Ihsanoğlu, Ekmeleddin. "Modern Islam." In *Science and Religion Around the World*, edited by John Hedley Brooke and Ronald L. Numbers, 148–74. Oxford: Oxford University Press, 2011.

Institute of Education Sciences: National Center for Education Statistics. "Science 2009: National Assessment of Educational Progress at Grades 4, 8, and 12." January 2011. Accessed August 2012. http://nces.ed.gov/nationsreportcard/pubs/main2009/2011451.asp.

Iqbāl, Muẓaffar. *Science and Islam.* New York: Greenwood, 2007.

Jefferson, Thomas. *Public Papers of Thomas Jefferson 1743–1826.* University of Virginia Library Electronic Text Center. http://etext.virginia.edu/toc/modeng/public/JefPapr.html

Jefferson, Thomas. *Report of the Commissioners for the University of Virginia.* Electronic Text Center, University of Virginia Library. http://etext.virginia.edu/toc/modeng/public/JefRock.html.

Johnson, Phillip E. *The Wedge of Truth: Splitting the Foundations of Naturalism.* Downers Grove, IL: InterVarsity Press, 2000.

Kaestle, Carl F. *Pillars of the Republican: Common Schools and American Society, 1780–1860.* New York: Hill and Wang, 1983.

Kahn, Ibrahim. "Three Views of Science in the Islamic World." In *God, Life, and the Cosmos: Christian and Islamic Perspectives,* edited by Ted Peters, Muzaffar Iqbāl, and Syed Normanul Haq, 43–75. Aldershot, UK: Ashgate, 2002.

Katz, Michael S. *A History of Compulsory Education Laws. Fastback Series, No. 75. Bicentennial Series.* Bloomington, IN: Phi Delta Kappa, 1976.

Kennedy, John F. "Science as a Guide of Public Policy." Address to the National Academy of Sciences, Washington, D.C. October 22, 1963.

Kessel, Alisa. "The New Parochialism?: Liberal education and the danger of teaching tolerance" 2012 unpublished manuscript.

Kevles, Daniel. *In the Name of Eugenics: Genetics and the Uses of Human Heredity.* Cambridge, MA: Harvard University Press, 1998.

King, David. *Astronomy in the Service of Islam.* Aldershot, UK: Variorum, 1993.

King, David. *Islamic Mathematical Astronomy.* London: Variorum, 1986.

Kitcher, Philip. *Abusing Science: The Case Against Creationism.* Cambridge and London: MIT Press, 1982.

Kitcher, Philip. *Living with Darwin: Evolution, Design, and the Future of Faith.* Oxford: Oxford University Press, 2006.

Kitcher, Philip. *Science, Truth, and Democracy.* Oxford: Oxford University Press, 2003.

Knight, Jack, and James Johnson. *The Priority of Democracy: Political Consequences of Pragmatism.* New York: Russell Sage Foundation, 2011.

Kurland, Philip B., and Ralph Lerner, eds. *The Founders' Constitution.* Chicago: University of Chicago Press, 1987. Accessed February 9, 2010. http://press-pubs.uchicago.edu/founders/.

Labov Jay B., and Barbara Kline Pope. "Understanding Our Audiences: The Design and Evolution of *Science, Evolution, and Creationism.*" *The American Society for Cell Biology Education—Life Sciences Education* 7 (Spring 2008): 20–24. Accesssed August 17, 2011. http://www.lifescied.org/content/7/1/20.full

Langer, Susanne K. *Philosophy in a New Key: A Study in the Symbolism of Reason, Rite, and Art.* Cambridge, MA: Harvard University Press, 1996.

Larson, Edward J. *Summer for the Gods: The Scopes Trial and America's Continuing Debate Over Science and Religion.* New York: Basic Books, 2006.

Lebo, Lauri. *The Devil in Dover: An Insider's Story of Dogma v. Darwin in Small-town America.* New York: New Press, 2008.

Levinson, Meira. *The Demands of a Liberal Education.* Oxford: Oxford University Press, 2002.

Liebell, Susan. "Lockean Switching: Imagination and the Production of Principles of Toleration." *Perspectives on Politics* 7, no. 4 (December 2009): 823–36.

Liell, Scott M. "Shaking the Foundation of Faith." *The New York Times,* November 18, 2005.

Lienesch, Michael. *In the Beginning: Fundamentalism, The Scopes Trial, and the Making of the Anti-evolution Movement.* Chapel Hill: University of North Carolina Press, 2007.

Lindberg, David C. "The Medieval Church Encounters the Classical Tradition: Saint Augustine, Roger Bacon, and The Handmaiden Metaphor." In *When Science and Christianity Meet,* edited by David C. Lindberg and Ronald L. Numbers, 7–32. Chicago: University of Chicago Press, 2008.

Lindberg, David C., and Ronald I. Numbers, eds. *God and Nature: Historical Essays on the Encounter between Science and Religion.* Berkeley: University of California Press, 1986.

Linder, Douglas O. "Famous Trials in History." University of Missouri-Kansas City School of Law. Accessed August 20, 2012. http://law2.umkc.edu/faculty/projects/ftrials/scopes/scopes2.htm.

Livingstone, David N. *Darwin's Forgotten Defenders.* Grand Rapids, MI: Eerdmans, 1987.

Locke John. *Some thoughts concerning education; and, Of the conduct of the understanding,* edited by Ruth Grant and Nathan Tarcov. Indianapolis: Hackett Pub. Co., 1996.

Locke, John. "Second Treatise." In *Two Treatises of Government,* edited by Peter Laslett. Cambridge: Cambridge University Press, 1988.

Lucas, J.R. "Wilberforce and Huxley: A Legendary Encounter." *The Historical Journal* 22, no. 2 (June 1979): 313–30.

Macedo, Stephen. *Diversity and Distrust: Civic Education in a Multicultural Democracy.* Cambridge, MA: Harvard University Press, 2003.

Macedo, Stephen. "Liberal Civic Education and Religious Fundamentalism: The Case of God v. John Rawls?" In *The Ethics of Citizenship: Liberal Democracy and Religious Convictions,* edited by J. Caleb Clinton, 93–117. Waco, TX: Baylor University Press, 2009.

MacMullen, Ian. *Faith in Schools? Autonomy, Citizenship, and Religious Education in the Liberal State.* Princeton: Princeton University Press, 2007.

Madison, James. "Journal of the Constitutional Convention of 1787." In *The Writings of James Madison, Comprising His Public Papers and His Private Correspondence, Including his Numerous Letters and Documents Now for the First Time Printed,* edited by Gaillard Hunt. New York: G.P. Putnam's Sons, 1900, vol. 3. Accessed September 10, 2012. http://oll.libertyfund.org/title/1935/118621/2395726.

Maimonedes. *Guide to the Perplexed* and "Letter on Astrology." In *A Maimonides Reader,* edited by Isadore Twersy. New York: Behrman House, 1972.

Mayell, Hillary. "Oldest Human Fossils Identified." *National Geographic News.* February 16, 2005. Accessed October 20, 2012. http://news.nationalgeographic.com/news/2005/02/0216_050216_omo.html.

McDougall, Walter. *The Heavens and the Earth: A Political History of the Space Age.* New York: Basic Books, 1985.

Mencken, H.L. *A Religious Orgy in Tennessee: A Reporter's Account of the Scopes Monkey Trial.* New York: Melville House 2006.

Mill, J.S. *On Liberty and Other Writings,* edited by Stefan Collini. Cambridge: Cambridge University Press, 1991.

Miller, Jon D., Eugenie C. Scott, and Shinji Okamoto. "Science Communication: Public Acceptance of Evolution." *Science* 313 (2006): 765–66.

Miller, Kenneth R. *Only a Theory: Evolution and the Battle for America's Soul.* New York: Viking, 2008.

Moore, James R. *The Post-Darwinian Controversies: A Study of the Protestant Struggle to Come to Terms with Darwin in Great Britain and America, 1870–1900.* Cambridge: Cambridge University Press, 1979.

Moore, James R. "Ronald Aylmer Fisher: A Faith Fit for Eugenics." In *Eminent Lives in Twentieth-Century Science & Religion,* edited by Nicholas A. Rupke, 103–38. New York: Peter Lang, 2007.

Morris, Henry. "The ICR Scientists." *Impact* 86 (August 1980).

Nagel, Thomas. "Public Education and Intelligent Design." *Philosophy & Public Affairs* 36, no. 2 (2008): 187–205.

National Academy of Sciences. *Evolution and Creationism: A View from the National Academy of Sciences,* 2nd Edition. Washington, D.C.: National Academy Press, 1999.

National Academy of Sciences. *Evolution in Hawaii: A Supplement to Teaching About Evolution and the Nature of Science.* Washington, D.C.: National Academies Press, 2004.

National Academy of Sciences. *Teaching About Evolution and the Nature of Science.* Washington, D.C.: National Academy Press, 1998.

National Academy of Sciences and Institute of Medicine. *Science, Evolution, and Creationism.* Washington, D.C.: National Academies Press, 2008.

National Academy of Sciences. "Evidence Supporting Biological Evolution." *Science and Creationism: A View from the National Academy of Sciences,* 2nd ed. Washington, D.C.: National Academies Press, 1999. Accessed January 27, 2013. http://www.nap.edu/catalog.php?record_id=6024.

Newman, John Henry. "Letter to J. Walker of Scarborough, May 22, 1868." In *The Letters and Diaries of John Henry Newman.* Oxford: Clarendon Press, 1973.

Newport, Frank. "Evolution, Creation, Intelligent Design." Gallup Poll, January 30, 2013. Accessed February 1, 2013. http://www.gallup.com/poll/21814/evolution-creationism-intelligent-design.aspx.

Newport, Frank. "In U.S., 46% Hold Creationist View of Human Origins." Gallup Poll, June 1, 2012. Accessed July 7, 2012. http://www.gallup.com/poll/155003/Hold-Creationist-View-Human-Origins.aspx?version=print.

Numbers, Ronald L. *The Creationists: From Scientific Creationism to Intelligent Design,* expanded edition. Cambridge, MA: Harvard University Press, 2006.

Nussbaum, Martha C. *Not for Profit: Why Democracy Needs the Humanities.* Princeton: Princeton University Press, 2010.

Nussbaum, Martha C. *Liberty of Conscience: In Defense of America's Tradition of Religious Equality.* New York: Basic Books, 2008.

Nussbaum, Martha C. *Frontiers of Justice: Disability, Nationality, Species Membership.* Cambridge, MA: Harvard University Press, 2007.

Nussbaum, Martha Cr. *Cultivating Humanity: A Classical Defense of Reform in Liberal Education.* Cambridge, MA: Harvard University Press, 1997.

O'Leary, Don. *Roman Catholicism and Modern Science: A History.* New York: Continuum, 2007.

Okin, Susan Moller. "Mistresses of Their Own Destiny: Group Rights, Gender, and Realistic Rights of Exit." In *Education and Citizenship in Liberal-Democratic Societies: Teaching for Cosmopolitan Values and Collective Identities,* edited by Kevin McDonough and Walter Feinberg, 325–50. Oxford: Oxford University Press, 2003.

Okin, Susan Moller. "Forty Acres and a Mule' for Women: Rawls and Feminism." *Politics, Philosophy & Economics* 4 (June 2005): 233–48.

Paine, Thomas. "The Age of Reason." In *The Writings of Thomas Paine,* vol. 4, edited by Moncure Daniel Conway. New York: G.P Putnam's Sons, 1894. Kindle edition.

Peacocke, Arthur. *Evolution: The Disguised Friend of Faith?* West Conshohocken, PA: Templeton Foundation Press 2004.

Peacocke, Arthur. *Creation of the World of Science.* Oxford: Oxford University Press, 2004.

Pennock, Robert T., ed. *Intelligent Design Creationism and Its Critics: Philosophical, Theological, and Scientific Perspectives*. Cambridge, MA and London: MIT Press, 2001.

Peterson, Merrill, ed. *Thomas Jefferson Writings*. New York: Library of America, 1984.

Pew Research Center for the People and the Press. "Public Opinion on Religion and Science in the United States." July 9, 2009. Accessed October 11, 2009. http://www.people-press.org/files/legacy-pdf/528.pdf.

Pigliucci, Massimo. *Nonsense on Stilts: How to Tell Science from Bunk*. Chicago: University of Chicago Press, 2010.

Plutzer, Eric, and Michael Berkman. "Evolution, Creationism, and the Teaching of Human Origins in Schools." *Public Opinion Quarterly* 72, no. 3 (Fall 2008): 540–53.

Polkinghorne, John. *Exploring Reality: The Intertwining of Science and Religion*. New Haven: Yale University Press, 2007.

Pollard, William G. *Chance And Providence; God's Action In A World Governed By Scientific*. New York: Scribner, 1958.

Post, Robert C. "Informed Consent to Abortion: A First Amendment Analysis of Compelled Physician Speech." Faculty Scholarship Series, 2007, Paper 170. Accessed January 3, 2013. http://digitalcommons.law.yale.edu/fss_papers/170.

Public Religion Research Institute. "PRRI/RNS Religion News Survey September 14–18, 2011." Accessed April 13, 2012. http://publicreligion.org/site/wp-content/uploads/2011/10/September-PRRI-RNS-Topline-Questionnaire-and-Survey-Methodology-.pdf.

Rawls, John. *A Theory of Justice,* revised edition. Cambridge, MA: Harvard University Press, 2005.

Rawls, John. "The Idea of Public Reason Revisited." *The University of Chicago Law Review* 64, no. 3 (Summer 1997): 765–807.

Reardon, Sara. "Bill Allowing Teachers to Challenge Evolution Passes Tennessee House." *Science Insider*, April 7, 2011. Accessed September 12, 2012. http://news.sciencemag.org/scienceinsider/2011/04/bill-allowing-teachers-to-challenge.html.

Reese, William J. *America's Public Schools: From the Common School to "No Child Left Behind."* Baltimore: The Johns Hopkins University Press, 2005.

Reich, Rob, and William S. Koski. "The State's Obligation to Provide Education: Adequate Education or Equal Education?" Unpublished manuscript, American Political Science Association, 2007.

Reich, Rob. *Bridging Liberalism and Multiculturalism in American Education*. Chicago: University of Chicago Press, 2002.

"Resolution 1982-D090," *The Archives of the Episcopal Church, Acts of Convention 1976–2009*. Accessed August 28, 2012. http://www.episcopalarchives.org/cgi-bin/acts/acts_resolution.pl?resolution=1982-D090.

"Resolution 2006-A129: Affirm Evolution and Science Education." *The Archives of the Episcopal Church, Acts of Convention 1976–2009*. Accessed August 28, 2012. http://www.episcopalarchives.org/cgi-bin/acts/acts_resolution-complete.pl?resolution=2006-A129.

Roach, John. "Intelligent design' in Tenn. schools?," *Cosmic Log on NBCNEWS.COM*, April 8, 2011. Accessed April 7, 2011. http://cosmiclog.msnbc.msn.com/_news/2011/04/08/6433919-intelligent-design-in-tenn-schools?lite.

Rogers, Melvin I. *The Undiscovered Dewey: Religion, Morality, and the Methods of Democracy*. New York: Columbia University Press, 2009.

Ruse, Michael. *Can a Darwinian Be a Christian? The Relationship between Science and Religion*. Cambridge, MA: Harvard University Press, 2001.

Ruse, Michael. *Darwin and Design: Does Evolution Have a Purpose?* Cambridge, MA: Harvard University Press, 2003.

Ruse, Michael. *The Darwinian Revolution: Science Red in Tooth and Claw.* Chicago: University of Chicago Press, 1979.

Ruse, Michael. *The Evolution-Creation Struggle.* Cambridge, MA: Harvard University Press, 2005.

Ruse, Michael. *Monad to Man: The Concept of Progress in Evolutionary Biology.* Cambridge, MA: Harvard University Press, 1996.

Samuleson, Norbert. "Judaism and Science." In *The Oxford Handbook of Religion and Science*, edited by Philip Clayton, 41–44. Oxford: Oxford University Press, 2006.

Sardar, Ziyauddin. *Explorations in Islamic Science.* London: Mansell, 1989.

Sardar, Ziyauddin. *Islamic Futures: The Shape of Ideas to Come.* London: Mansell, 1985.

Schattner, Elaine. "Do Abortions Cause Breast Cancer? The Shaky Science Behind Kansas' House Abortion Act." *Slate*, May 23, 2012. Accessed January 3, 2013. http://www.slate.com/articles/health_and_science/medical_examiner/2012/05/do_abortions_cause_breast_cancer_kansas_state_house_abortion_act_invokes_shaky_science_for_political_gain_.html.

Scheiber, Noam. "Known Unknowns." *The New York Times Book Review.* Sunday November 4, 2012.

Schroeder, Gerald L. *The Science of God: The Convergence of Scientific and Biblical Wisdom.* New York: Broadway Books, 1998.

Silver, Nate. *The Signal and the Noise: Why So Many Predictions Fail—but Some Don't.* New York: Penguin, 2012.

Slack, Gordy. *The Battle over the Meaning of Everything: Evolution, Intelligent Design, and a School Board in Dover, PA.* San Francisco: John Wiley & Sons, 2007.

Slifkin, Natan. "A General Response to the Charge of Heresy." Accessed August 2012. http://www.zootorah.com/controversy/scienceresponse.html.

Slifkin, Natan. *The Challenge of Creation: Judaism's Encounter with Science, Cosmology and Evolution.* Yashar Books, 2006.

Slifkin, Natan. *The Science of Torah: The Reflections of Torah in the Laws of Science, the Creation of the Universe, and the Development of Life.* Southfield, MI: Targum, 2001.

Smith, Rogers M. *Civic Ideals.* New Haven: Yale University Press, 1997.

Snyder, Thomas D., and Alexandra G. Tan. *Digest of Education Statistics, 2004.* Washington, D.C.: U.S. Government Printing Office, 2005.

Spinner-Halev, Jeff. *The Boundaries of Citizenship: Race, Ethnicity and Nationality in the Liberal State.* Baltimore: The Johns Hopkins University Press, 1994.

"The State of State Science Standards." Thomas Fordham Institute. January 31, 2012. Accessed August 2012. http://www.edexcellence.net/publications/the-state-of-state-science-standards-2012.html.

Stolzenberg, Nomi Maya. " 'He drew a circle that shut me out': Assimilation, Indoctrination, and the Paradox of a Liberal Education." *Harvard Law Review* 106 no. 3 (January 1993), 581–667.

Thompson, Helen. "Tennessee 'Monkey Bill' Becomes Law." *Nature: International Weekly Journal of Science*, April 11, 2012. Accessed April 7, 2011. http://www.nature.com/news/tennessee-monkey-bill-becomes-law-1.10423.

Thomson, Keith. *A Passion for Nature: Thomas Jefferson and Natural History.* Monticello: Thomas Jefferson Foundation, 2008.

"U.S. Rep. Paul Broun: Evolution a Lie 'from the pit of hell.' " *Los Angeles Times.* October 7, 2012. Accessed October 20, 2012. http://articles.latimes.com/2012/oct/07/nation/la-na-nn-paul-broun-evolution-hell-20121007.

Welter, Rush. *Popular Education and Democratic Thought in America.* New York and London: Columbia University Press, 1962.

Wing, Nick. "Tennessee Evolution Bill Becomes Law After Governor Bill Haslam Declines To Act." *The Huffington Post.* Accessed August 2012. http://www.huff ingtonpost.com/2012/04/10/tennessee-evolution-bill-haslam_n_1416015. html.

Witham, Larry. "Many Scientists See God's Hand in Evolution." *Reports of the National Center for Science Education* 17, no. 6 (November–December 1997): 33. Accessed September 19, 2012. http://ncse.com/rncse/17/6/many-scientists-see-gods-hand-evolution.

Wolin, Sheldon S. "Political Theory as a Vocation." *The American Political Science Review* 63, no. 4 (Dec. 1969): 1062–82.

Wood, Paul, ed. *Science and Dissent in England, 1688–1945.* Aldershot, UK: Ashgate, 2004.

Yahya, Hârun. *The Evolution Deceit: The Scientific Collapse of Darwinism and Its Ideological Background.* London: Ta-Ha, 1999.

Court Cases

Abington School District v. Schempp. 374 U.S. 203 (1963).

Abrams v. U.S. 250 U.S. 616 (1919).

Ambach v. Norwick. 441 U.S. 68 (1979).

Brief for 72 Nobel Laureates, 17 State Academies of Science, and 7 other Scientific Organizations as Amicus Curiae, Edwards v. Aguillard, 482 U.S 578 (1987).

Brief for National Academy of Sciences as Amicus Curiae, Edwards v. Aguillard, 482 U.S 578 (1987). Accessed April 2010. http://www.soc.umn.edu/~samaha/cases/edwards_v_aguillard_NAC.html.

Brown v. Board of Education. 347 U.S. 483 (1954).

Campaign for Fiscal Equity v. State. 100 N.Y.2d 893 at 951 (2003), 801 N.E.2d 326.

County of Allegheny v. American Civil Liberties Union. 492 U.S. 573 (1989).

Daniels v. Waters. 515 F. 2d 485 (6th Cir. 1975).

Don Aguillard, et al. v. Edwin Edwards, et.al. 765 F.2d 1251 (5th Cir. 1985).

Edwards v. Aguillard. 482 U.S. 578 (1987).

Epperson v. Arkansas, 393 U.S. 97 (1968).

John Scopes v. The State. 154 Tenn. 105, 289 S.W. 363 (Tenn. 1927).

Lamont v. Postmaster General. 381 U.S. 308 (1965).

Lemon v. Kurtzman. 403 U.S. 602 (1971).

Lynch v. Donnelly. 465 U.S. 668 (1984).

McLean v. Arkansas. 529 F. Supp. at 1258.

Meyer v. Nebraska. 262 U.S. 390 (1923).

Mozert v. Hawkins County Board of Education. 827 F.2d 1058 (6th Cir. 1987).

Plyler v. Doe. 457 U.S. 202 (1982).

Reynolds v. Sims. 377 U.S. 533 (1964).

San Antonio Independent School Dist. v. Rodriguez. 411 U.S. 1, 35 (1973).

Schenck v. United States. 249 U.S. 47 (1919).

Selman v. Cobb County School District. 449 F.3d 1320 (11th Cir. 2006).

Tammy Kitzmiller, et al. v. Dover Area School District, et al., 400 F. Supp. 2d 707 (M.D. Pa 2005). Accessed March 2010. http://www.pamd.uscourts.gov/kitzmiller/kitzmiller_342.pdf.

William v. Rhodes. 393 U.S. 23 (1968).

Wisconsin v. Yoder. 406 U.S. 205 (1972).

Statutes

Tennessee Critical Inquiry Law. Amendment No.1 to HB 368/SB 893. Accessed July 23, 2012. http://www.capitol.tn.gov/Bills/107/Amend/SA0901.pdf.

Texas Right to Know Act, 2003. Health and Safety Code, Sec. 171.012. Accessed July 23, 2012. http://www.statutes.legis.state.tx.us/Docs/HS/htm/HS.171.htm.

Pennsylvania Code of Professional Practice and Conduct for Educators. §235.10(2).

Index